The Klamath Knot

Explorations of Myth and Evolution

David Rains Wallace

Illustrations by Karin Wikström

Twentieth Anniversary Edition

UNIVERSITY OF CALIFORNIA PRESS

Berkeley   Los Angeles   London

LIBRARY OF CONGRESS CATALOGING-IN-PUBLICATION DATA

Wallace, David Rains, 1945–
    The Klamath knot : explorations of myth and evolution / David Rains Wallace ;
illustrations by Karin Wikström.—20th anniversary ed.
      p. cm.
    Originally published: 1st ed. San Francisco : Sierra Club Books, c1983.
    ISBN 0-520-23659-9
    1. Natural history—Klamath Mountains (Calif. and Or.)   2. Evolution (Biology)
3. Klamath Mountains (Calif. and Or.)   I. Title.

QH105.C2 W344 2003
508.795'2—dc21   2002038727

Printed in the United States of America
10 09 08 07 06 05 04 03 02
10 9 8 7 6 5 4 3 2 1

The paper used in this publication meets the minimum requirements of ANSI/NISO
Z39.48-1992 (R 1997) (*Permanence of Paper*). ∞

TO JON BECKMANN

CHAPTER ONE

TRACKS IN THE WILDERNESS

*Mountains lie all about, with many difficult turns leading here
and there. The trails run up and down; we are martyred
by obstructing rocks. No matter how well we keep the path, if we
miss one single step, we shall never know safe return.
But whoever has the good fortune to penetrate that wilderness,
for his labors will gain a beatific reward. The wilderness abounds
in whatsoever the ear desires to hear, whatsoever pleases the eye.*

Gottfried von Strassburg
Tristan

I FIRST WENT to the Klamath Mountains in 1969, with a fifteen-
dollar sleeping bag, a canvas Boy Scout knapsack, some canned
goods, and a peculiarly unreliable U.S. Forest Service recreation
map. I didn't get far into the woods. I spent most of the time
nervously driving my secondhand Volkswagen over hideously rocky
logging roads, searching for hiking trails that seemed to have been
swallowed by the logging roads, though they were marked on my
map. In effect, I was walled into the Klamath River gorge by steep,
brushy slopes, piles of logging slash, and my own ignorance. The
wilderness I was vaguely seeking, the high Siskiyous, remained a
paper expanse of Forest Service green marked with tantalizingly
picturesque place names — Cyclone Gap, No Man's Creek, Dark
Canyon.

Still, I came away from these days of frustration with two strong impressions. The first struck me at twilight as I looked west toward the heavily forested Siskiyou ridgeline. It was one of the clear, almost colorless sunsets typical of the Klamath Mountains in summer, the cloudless sky turning a dusky orange-red that merely deepened the heavy forest green of the ridges. A slight wind pulled from the east, otherwise the landscape was motionless. Despite the picture-postcard aspect of its pines and peaks, it was the strangest landscape I had ever seen.

The ridges were not particularly high or craggy, rather a succession of steep, pyramidal shapes that marched almost geometrically into blue distance. The big conifers that crowned them enhanced an impression of regularity, almost of discipline. There was a tension in the ridges that departed radically from conventional notions of the irregularity and relaxation of wide open spaces. It was almost an *attention*. I felt the hair stand up on the backs of my arms and legs. The faint sibilance of wind in pine needles called attention to a quiet so intense that I was reluctant to move, like a grouse chick crouched on the forest floor.

The ridges seemed not only vigilant, but reticent, as though hidden within them might be the most extraordinary things. Perhaps this impression was colored by my awareness that I was looking toward the Bluff Creek drainage, where giant, humanoid footprints had been found in the dust of a road-building project in the early 1960s. The pyramidal ridges seemed to say "mystery" to my mind in the way that the shape or color of a parent bird's bill says "food" to its nestlings. Pyramids have a way of doing that, as evidenced by the lasting fascination of certain Egyptian tombs. The Siskiyou ridges might have been the vegetated remnants of some prehistoric city, vast beyond comprehension. They did not seem altogether natural, at least, not with the insensate simplicity often associated with nature.

Nightfall deepened the impression of attentiveness. The trees looked much taller against the stars, which came out in a profusion I'd never seen before. The ridges were outlined clearly against the starry horizon. As I sat feeding a small fire against the chill, I noticed that a grove of trees nearby had begun to glow with a cold but surprisingly bright light. The light spread down to the canyon floor and struck pale highlights on the rocks, as though some presence were approaching through the trees. Then something caught my eye in the opposite direction, down the canyon. I turned my head and saw a light so brilliantly white I had to shield my eyes from it, as from a headlight. The moon had just risen above the ridges and was casting its beams into the canyon.

The subject of my second impression was more mundane. The following afternoon I was trudging down a dusty road, having again failed to find a mapped trail, when I chanced upon a fresh pile of bear scat. The dust clearly showed how the bear had ambled to the road's center, defecated a heap of berry seeds, root fibers, and rodent hairs, then wandered downhill. I had never seen a wild bear, and I think this was the first clear evidence I'd seen of one's immediate presence. It was somehow dizzying to come upon. It struck me with something that I had known but never felt: bears precede our perception of them, they came before national forests, before the *word* forest — black, shaggy beings emergent from millions of forested years without benefit of manufacture or legislation.

I think this vertigo overthrew for the first time my unconscious childhood assumption that the world had been made by my parents or some other human authority. The feeling recalled dreams I had experienced in a time of postadolescent anxiety, dreams in which I floated above great depths of air or water so bright and clear that animals and plants beneath me glowed with colors beyond my capacity to describe, colors without names. They had been enthralling dreams, but a little frightening, since there was always the possibility that I might fall or sink into the depths. But this had never happened. Instead, as the dreams continued over several years, the beings in the depths began to rise toward me, with increasing surges of power, until (while sleeping on the floor of a friend's apartment in Manhattan on a sticky summer night) I found myself at the apex of a maelstrom of creatures — fish, snakes, seals, blue horses with wings and scaly, serpentine, fishtails. Far from being frightened, I had felt buoyed up by this fountain of life, reassured; and after a few more such dreams, my anxieties had quieted. The dreams had gone away.

There was little chance of my floating through the air in the Klamath Mountains, but my impression of the bear's deep ancestry was not the less strong for being a waking, mundane one. In a way it was stronger. Vast as the spaces of my dreams had been, I felt I was glimpsing much greater spaces, which my dreams had only reflected. I was seeing a bear as it really is, not only a black animal in a forest, but part of a long wave of black animals surging upward from depths of time imperceptible to normal senses. I felt as though I had seen in four dimensions for a moment, as though some nascent or atrophied sense organ had given me a twinge.

Ten years passed before I went back to the Siskiyous. During that time I walked into a number of wild places, and acquired what I thought was a fair knowledge of western mountain wilderness: of the climb from chaparral or sagebrush in the Upper Sonoran Zone; through Douglas fir, ponderosa pine, and white fir in the Transition Zone; past lodgepole pine, red fir, or Engelmann spruce in the Canadian Zone; to stunted whitebark pines and heather in the Alpine Zone. I went to a few places where there were still grizzly bear tracks as well as black bear tracks. So I didn't really expect to find much that was new when I started up the Clear Creek trail into the northern part of the high Siskiyou in June of 1979 with my down sleeping bag, gas stove, contour maps, and other sophistications. But the Siskiyous still had some things to show me.

I knew the Siskiyous are among the richest botanical areas of the West, and I soon saw evidence of this as I followed Clear Creek upstream. Tributary ravines contained so much blossoming azalea that the forest often smelled like a roomful of fancy women, and rhododendrons were in flower on one flat bench. There were more orchids than I'd seen anywhere. California lady's slippers hung over one rivulet like tiny Japanese lanterns dipped in honey, and I found three species of coralroot, red and orange orchids that have no green leaves, lacking chlorophyll. Farther up the trail, where snow had melted recently, pink calypso orchids had just burst through the pine duff.

The forest that overshadowed these flowers was the most diverse I'd seen west of the Mississippi. Besides the Douglas fir, tan oak, madrone, golden chinquapin, and goldencup oak I had expected just east of the coastal crest, I found ponderosa pine, Jeffrey pine, sugar pine, western white pine, knobcone pine, and incense cedar. Moist ravines were full of Port Orford cedar, a lacy-foliaged tree with fluted bark like a redwood's. The diversity became confusing; it seemed I had to consult my tree field guide every few minutes.

As I climbed higher, I kept expecting this unwonted diversity to sort itself out into the usual altitudinal zones, waiting for white fir, ponderosa pine, and incense cedar to close ranks against the confusion. But it didn't happen. Douglas fir kept playing its polymorphous tricks, its foliage sometimes resembling the flattened needles of white fir, sometimes dangling like the branches of weeping spruce. I got a stiff neck looking up to see if cones hung downward, denoting Douglas fir, or stood upright, denoting white fir (or perhaps silver fir, grand fir, or noble fir, three other species found in the Klamaths).

Broad-leaved madrone and tan oak disappeared obligingly after I reached a certain altitude, but then new species appeared. I found western yew, a sturdy little tree resembling a miniature redwood, and Sadler's oak, another small tree whose serrated leaves reminded me of the chestnut oaks I'd known in the Midwest. I passed a grove of lodgepole pines, and these austere trees, which typically grow on bleak, windswept terrain, looked out of place in all the effulgent variety. The trees were sorted out somewhat according to soil conditions, but these distinctions were patchy and vague, offering cold comfort to my organizing instincts.

After two days of walking, I stood on the slopes of Preston Peak, which is 7,309 feet above sea level at its summit but seems higher as it thrusts abruptly above the forested ridges. I was surprised, on looking around at the snow-stunted trees on the glacial moraine where I stood, to find that they were the same species that had accompanied me from the Klamath River: Douglas fir, ponderosa pine, incense cedar, western yew, Sadler's oak, white fir. Even goldencup oak, golden chinquapin, and bay laurel grew there at about 5,000 feet, albeit in shrubby form.

Clearly, there was something odd about the Siskiyou forest. For so many species to grow all over a mountain range simply doesn't conform to respectable western life-zone patterns. It is more like some untidy temperate deciduous forest or tropical rainforest, species promiscuously tumbled together without regard for ecological proprieties.

The high Siskiyou forest is a rare remnant of a much lusher past. Fossils of trees almost identical to those of the Siskiyous have been dug from twelve-million-year-old, Pliocene epoch sediments in what are now the deserts of Idaho and eastern Oregon. Fossils of trees not at all unlike Siskiyou species have been found in *forty*-million-year-old sediments in Alaska. In that epoch, the Eocene, a temperate forest surpassing any living today covered the northern half of this continent from coast to coast. Redwoods, pines, firs, and cedars grew with hickories, beeches, magnolias, and other hardwoods not found within a thousand miles of the Pacific Ocean today, and with ginkgoes, dawn redwoods, and other trees that don't even grow naturally in North America anymore. It is hard to imagine such a forest: it sounds like poets' descriptions of Eden. After the Eocene, though, the climate became cooler and drier; and this gradually drove the forest southward, and split it in half. Deciduous hardwoods migrated southeast, where the summer rain they needed was still available, while many conifers migrated southwest to cover the growing Rocky Mountain and Pacific Coast ranges. Ginkgoes and dawn redwoods fell by the wayside during this "long march," which has resulted in our present, relatively impoverished forests, where trees that once grew together are separated by wide prairies and plains.

There is still one area west of the Rockies, however, where rainfall and temperatures approximate the benign Eocene environment: the inner coastal ranges of southwest Oregon and northwest California, the Klamath Mountains. In the Klamaths, winters are mild enough and summers moist enough for species to grow together that elsewhere are segregated by altitude or latitude. Several species that once grew throughout the West now survive only in the Klamaths. Perched on my Siskiyou eminence, I again felt suspended over great gulfs of time. The stunted little trees and their giant relatives on the lower slopes were not a mere oddity forest where ill-assorted species came together in a meaningless jumble. They were in a sense the ancestors of all western forests, the rich gene pool from which the less varied, modern conifer forests have marched out to conquer forbidding heights from Montana to New Mexico. Looking out over the pyramidal Siskiyou ridges, I was seeing a community of trees at least forty million years old.

Later that day something hair-raising happened. There were still some patches of snow, and I had walked across one on the way to my campsite. After dinner I wandered back past that patch and found, punched deeply into each of my vibram-soled footprints, the tracks of a large bear. It probably had been foraging in Rattlesnake Meadow, heard me coming, and took the trail downhill to escape my intrusion. A simple coincidence, but it caused a sudden feeling of emptiness at the pit of my stomach, as though I were riding a fast elevator. It seemed the lesson begun ten years before was proceeding: from a realization that the world is much greater and older than normal human perception of it, to a reminder that the human is a participant as well as a perceiver in the ancient continuum of bears and forests. I was used to walking in bear tracks by this time; it was instructive to find that a bear also could walk in mine.

The Siskiyous weren't through with me. I got sick the next day for some reason, probably fatigue. I'd been living in the Midwest for three years and had grown unaccustomed to running around on mountains. It was thought-provoking to lie in the wilderness that night with the suspicion that I might have been about to have a heart attack. I had many sleepless hours to wonder why I kept going to places like the Siskiyous when so many civilized places were so much easier to get to. I'm not all *that* crazy about exercise. Wilderness areas are certainly among the most beautiful places on the planet, but I wonder if this alone is enough to explain the fascination many people feel for them, or the difficulties and real suffering they endure to reach them. I thought of Audubon, feverish and vomiting from tainted turkey meat in the trackless Ohio forest; Thoreau dragging his tuberculosis to the Minnesota frontier; Muir stumbling with frostbite across Mount Shasta's glaciers. I may have been delirious: my mind started reeling through history — tribal youths starving on mountaintops for totem visions, Taoist sages living on nettles and mushrooms in Chinese caves, Hebrew prophets eating locusts and wild honey on the Sinai peninsula, elderly Brahmins leaving comfortable estates to wander the Bengali jungle.

I wondered if my motives for going into wilderness might be more obscure, and more profound, than I had realized. While part of me was going into the mountains seeking the pleasures of exercise, self-reliance, accomplishment, and natural history, it seemed that another part was looking for things of which I had only a vague conscious awareness, as though a remote mountain or desert releases some innate human behavior, a kind of instinctive predilection for the mysterious.

So many major structures of belief have arisen at least in part from experiences in wilderness. This was to be expected with the oldest structures, such as animism and shamanism, since the entire world outside a Paleolithic camp was wilderness. But why should all the major religions of the modern world include a crucial encounter with wilderness — Moses, Jesus, and Mohammed in the desert mountains, Siddhartha in the jungle? And why should the predominant modern view of the origin and development of life have arisen from the five-year wilderness voyage of a Victorian amateur naturalist named Charles Darwin? There evidently is more to wilderness than meets the eye — more than water, timber, minerals, the materials of physical civilized existence. Somehow there are mental trees, streams, and rocks — psychic raw materials from which every age has cut, dammed, or quarried an invisible civilization — an imaginative world of origins and meanings — what one might call a mythology.

Placing Darwin in the tradition of Moses and Jesus may seem heresy from both the Judeo-Christian and scientific viewpoints, but I think the roles played by the three figures have been similar. They wrenched their respective cultures out of a complacency that amounted to self-worship and thrust them in new directions that (if not always entirely beneficial) enlarged the human perspective. Moses forced his society to accept a unifying law; Jesus forced his to accept the unity of all humanity; Darwin forced his to accept the unity of all life. I doubt whether any of the three would have been able to influence his society so strongly if he had not been fortified by a season in the wilderness.

Both religion and science are mythologies, in the sense that each provides the individual with an account of the origins and meanings of life. It seems to me irrelevant, in this mythological sense, whether such accounts are fact or fiction. They need only provide their believers with a workable key to life, an invisible world of origins and meanings to help them make sense of an often confusing, sometimes frightening, physical world. As I lay sick in the Siskiyou night, I was comforted by my thoughts of origins and meanings, as though my existence, weakened and isolated, depended on keeping an invisible world alive in my mind. My fear at being sick and alone magnified a process that goes on all the time in wilderness — when I saw the Siskiyou ridgeline, the unimaginably venerable forest, the bear tracks in mine. Wilderness generates mythological thinking; it leads the mind back to stories of origins and meanings, to imagining the world's creation. Physical wilderness may have shrunk vastly from

Moses' to Darwin's time, but the growth of mythic wilderness has been greater — from the seven days' creation of the Bible to the over four billion years of a precivilized, wilderness earth in evolutionary theory. As it shrinks before us, wilderness expands around us.

I can't find an early mythology that held that civilization came before wilderness. Universally, the gods created people out of raw matter, then gave them the tools of human life. Only in the very modern flying-saucer cults is there the idea that humans did not originate on wilderness earth, but were dropped here by a civilization from elsewhere. And even that extraterrestrial civilization presumably would have had wild beginnings. The presence of wilderness — of various lonely expanses of sky, water, rock, and soil — in so many creation myths supports the evolutionary idea that humanity did in fact arise from a thoroughly wild planet. Whether the myths originated simply from the fact that primitive people were surrounded by wilderness all their lives or from some racial memory of wilderness at the beginning of the mind will be an unanswerable question until we understand the mind better. But the powerful resonance of wilderness in my own mind — never having seen a wilderness or really understood what one was until I was in my twenties — leads me to suspect there is some kind of genetic circuit that lights up when a suburban animal is set down before a virgin forest.

If wilderness generates mythology, it also is shaped by it. Wilderness after Darwin, the wilderness I walk into, is very different from that of Moses or Jesus. Evolution is the great myth of modern times. There has not been such a compelling new one for thousands of years, and even those who would discredit it are subtly bound by it, forced to seek evidence of the immutability of species in the very fossil record that so powerfully illustrates that species evolve and become extinct. And, although the truth or falsehood of evolution is irrelevant to its mythological function, there is no denying that it has greater scope, intricacy, and coherence than older myths, just as modern civilization is larger and more complex than its predecessors. Older myths see the earth's history as a matter of thousands of years; evolution sees it as a matter of billions. Older myths see life's creation as a matter of days, often by some rational intelligence; evolution sees the creation of new living beings as never ending, except in some hypothetical limit of cosmic time — a bizarre and unlikely (from the common-sense viewpoint) process of random molecular change whereby beings invisible to the naked eye grow with excruciating slowness into trees and people, which themselves eventually will grow by the same random process into things unimaginable by the human mind.

As with all new myths, evolution has grown from fundamental changes in human circumstances. Its roots lie deep in the decay of feudalism's reliance on the authority of received wisdom, and in the ferment of the pragmatic, skeptical, mercantile classes, who were open to a myth of continual change because, to them, change meant profit and improvement, not simply a falling away from Biblical or classical ideals. People have been finding fossils since the dawn of history, but they saw them as giants or monsters destroyed by the gods, not as inhabitants of natural landscapes ancestral to, though quite different from, the living world. It required the close observations of early geologists (who were looking for ways to improve agriculture and mining) to see that the bones were deposited in strata that became older, and more unlike the living landscape, the deeper they dug. The basic tenets of evolution were developed long before Darwin: that rocks relate the earth's history at least as reliably as scriptures; that present life is descended from the petrified bones and stems in the rocks.

Darwin's contribution to the science, and the myth, of evolution was the idea of natural selection and the struggle for existence. Darwin didn't prove that evolution occurs — the rocks did that — he showed *how* it occurs. Organisms evolve because population grows faster than food supply, and individuals better fitted to survive are more likely to leave offspring. Genetic traits that fit them for survival are thus favored, naturally selected, in the population as a whole, and the species eventually takes on the characteristics of the favored individuals. Natural selection explains why the trees growing in the Klamath Mountains now are not quite the same as those that grew in Alaska forty million years ago.

Darwin is not evolution's ultimate authority any more than he was its discoverer. Evolution has evolved, as all myths must. Past myths are psychic fossils. Darwin's natural selection doesn't explain how the genetic traits that imbued their possessors with superior fitness came into being. That required Mendel's work in genetics and the resultant concept of mutation, the idea that genes can change and thus produce entirely new characteristics in an organism. Furthermore, natural selection and gene mutation didn't explain how life arose in the first place, and they didn't fully explain how things as complicated as flowers evolved from things as relatively simple as algae. The Klamath Mountains are full of things that natural selection and mutation don't entirely explain. They explain why living oaks are different from fossil oaks, but they don't explain how oaks evolved from more primitive plants such as conifers. They operate too gradually to completely explain such enormous changes. Other concepts have arisen to try to account for the great leaps that life has taken: symbiosis, preadaptation, neoteny. These concepts are as picturesque as the most bizarre primitive lore, and in some ways just as mysterious.

Evolution has much in common with older myths. It tends to be cyclical, with successive worlds created and destroyed in satisfyingly catastrophic and mysterious ways. It is filled with colorful, well-loved characters. (There was a period in my childhood when I liked dinosaurs better than anything else in the world.) It is a handy way for older people to explain to younger ones how things became the way they are. (Given the instinctive curiosity of the young primate, this is no minor advantage.)

In some ways evolution has so far proved inferior to older myths. It lacks a clear ethical dimension, as evidenced by its misuse in brutal dogmas such as racism and social Darwinism. Nobody has figured out, as yet, ways to make concepts such as natural selection and mutation encourage people to be good, which is something older myths tended to do. Of course, older myths have been ethically misused too.

Evolution doesn't view earth's history as a conflict between good and evil. It does essentially view it as a conflict between life and death, between increased organization and more efficient energy use on the part of life, and an opposing tendency of nonliving matter to become disorganized and lose energy — entropy. But evolution doesn't see life and death as simple adversaries: life as good and death as evil. Life cannot triumph over death in evolution. They don't fight to win. As with some of the oldest myths, wherein the natural dualities of light and darkness, sun and moon, male and female, performed an eternal, amoral dance of opposites, evolutionary life and death are interdependent: two halves of the world. Evolution would be impossible if organisms did not die. Immortal organisms would never surrender the planet to their descendents, and thus the natural selection and mutation by which sexual reproduction changes organisms couldn't work. Many more early deaths than long lives are required for evolution to function.

The centrality and indispensability of death in evolution can make it seem horrible, like some bloody sacrificial fertility cult. Evolution can seem a throwback to a very savage view of life. It's not as simple as that, though. Evolution is not only a battle of numbers wherein the fit survive and the rest get dragged out by the heels. There are other ways for life to evolve besides competition. The conventional Darwinian picture of an apparently peaceful landscape which underneath is a seething battle for survival is after all a picture, an artifact superimposed on physical reality. To the pre-Darwinian senses, a peaceful landscape is just that. Landscapes have evolved from cooperation among organisms as well as from competition. If it seems anthropomorphic to speak of the cooperation between trees and insects, is it any less so to speak of their competing?

Evolution also lacks something of the immediacy of older myths, wherin the world's creation and other significant events were thought to have occurred at familiar, still-existing places, so that mythical events were a part of everyday experience and could seem virtually contemporary. Evolution largely has been presented as a thing of the distant past, its main references to the present — fossils, petrified bones and footprints — having little significance to the average person until interpreted by the specialist. It is not precise enough as a myth (and may never be precise enough even as a science) to allow people to look at some familiar landmark and think: This is where the *Brontosaurus* made its last stand against the forces that exterminated it. Nobody knows exactly what the forces that exterminated the *Brontosaurus* were, for one thing, and the geological nature of the planet does not lend itself to permanent landmarks. Perhaps one of the things that repel many people about evolution is this remoteness, this apparent distance from human life. It can seem to belittle and mock the living.

I don't think the remoteness is inherent in evolution, though. Evolution is fully as operative today as it was in the *Brontosaurus*'s time, even though it operates so subtly as to be perceptible only to the informed eye. More important, a surprisingly large number of the actors in various evolutionary dramas of mythic proportions are still with us, in the flesh as well as in the rocks. I don't know if the ancient Greeks actually saw nymphs and satyrs in their woods. I do know that there is a distinct possibility that modern people can see dinosaurs in *their* woods, and that there is no doubt that they can see creatures virtually identical to those that first populated the land a half-billion years ago. Every place on earth contains a treasury of evolutionary stories in its living animals and plants, for each is populated by a continuum of organisms that mimics the entire history of life, from the first cells to form in primal ooze, to the teeming invertebrates and fish of Paleozoic waters, to the first amphibians and insects on land, to the rich and diverse world of the dinosaurs — when redwoods grew in Greenland — to the drying, cooling world of the great mammals, our world.

The Klamath Mountains are an exceptionally rich storehouse of evolutionary stories, one of the rare places where past and present have not been severed as sharply as in most of North America, where glaciation, desertification, urbanization, and other ecological upheavals have been muted by a combination of rugged terrain and relatively benign climate. Klamath rocks are older than those of the California and Oregon coast ranges to the south and north or those of the Cascades to the east. They are more intricately and tortuously folded, faulted, and upthrust, forming a knot of jagged peaks and steep gorges less modified by civilization than other areas, even though they are only a day's drive from large cities. The Klamaths are not even very high as mountains go, with no peaks over ten thousand feet.

The relatively low elevation of the Klamaths, compared to the Cascades or Sierra Nevada, has caused them to be overlooked. Naturalists often say that the Klamaths are a combination of Sierra Nevada and Cascades ecosystems because the Klamaths contain species found in both other wilderness regions. This is a little like saying that a person is a combination of his brother and sister because he shares genes with both siblings. The Klamaths have a character of their own, although not perhaps as ingratiating a character as the graceful volcanic cones of the Cascades or the clean alpine country of the Sierra. There is something wizened about the Klamaths. Their canyons do not have sparkling granite walls and wide river meadows as do the U-shaped, glaciated canyons of the

Sierra. Klamath canyons are preglacial, and uncompromisingly V-shaped. They've never been scoured into spaciousness by the ice flows. They seem to drop down forever, slope after forest-smothered slope, to straitened, boulder-strewn bottoms so noisy with waters and shadowed by vegetation that they may bring startling dreams and uneasy thoughts to campers.

Early explorers were stymied by these canyons. In 1828 Jedediah Smith and his party of fur trappers gave up in despair when they tried to follow the Klamath River upstream from its confluence with the Trinity River. The terrain was too rugged even for those mountain men, who had walked from Oregon to Los Angeles in search of beaver. They didn't find many beaver in Klamath Mountain rivers, which are generally too rocky and turbulent even for those ingenious rodents. The fur trappers called the Klamaths "backward," a pretty definitive judgment coming from backwoodsmen who crossed the Sierra and Cascades, not to mention the Rockies, a half-century before the railroads.

More than any other wild region I've known, the Klamaths have a venerable quality which is not synonymous with "pristine," "unspoiled," or other adjectives commonly applied to natural areas. Certainly, the Klamaths are as unpolluted as any American place these days. But these adjectives imply something of the smoothness and plumpness of youth, whereas the Klamaths are marked by the wrinkles and leanness of great age. Although their peaks and high plateaus have been marked by glaciers, they are at heart preglacial mountains, with elements of flora and fauna that reach back farther into the past than any place west of the Mississippi River. The Klamaths seem so old, in fact, that I'd call them a grandparent of the Sierra and Cascades instead of a sibling.

This venerable quality is strongest in the region's National Forest wilderness areas: the Rogue River gorge and the jumbled red humps of the Kalmiopsis to the north, the jagged peaks of the high Siskiyous and Red Buttes, the huge massifs of the Marble Mountains and Salmon-Trinity Alps, the gentle but hulking summits of the Yolla Bollys to the south. (The Yolla Bollys aren't entirely within the Klamath Mountain geological province, but I include them because they're ecologically linked to the other ranges.) Wilderness in the Klamaths is still dwindling from logging and other developments, as it was when I found hiking trails so elusive in 1969, but I hope enough will eventually be protected to assure they will remain an outstanding vantage point into what I perceived during my first visit as the fourth dimension of life.

With Klamath Mountain wilderness as a vantage, then, this book will try to see evolution not as an edifice of petrified stems and bones, but as a living continuum still linked with its past. I will try to peer into time's depths as I peered into the depths of my dreams. The place of the human in evolution is not unlike the dreamer's in the dream. We drift on the surface of a vast gulf of time in which we fitfully perceive the creatures below us, and we may feel afraid at our suspended position or reassured by the presence of other drifters (as Alice was comforted in Wonderland by the little animals that floated with her in the sea of her tears). Whatever we feel, dreamers and creatures both are caught in time's current, which indifferently carries the most primitive and advanced of organisms, and which sometimes abandons apparent paragons of development, such as the dinosaurs, for reasons that remain obscure.

Such obscurities may seem to doom the enterprise from the start. Evolution is almost as rudimentary a science as it is a myth. The fossil, the rock upon which the entire edifice is built, is a scanty and unreliable phenomenon. An estimated 1 to 10 percent of all species that ever lived have left fossils, and only an estimated 1 to 10 percent of *those* will ever be found. The origins of major organisms such as flowering plants, protozoans, and frogs still remain obscured by a lack of transitional fossils. A species such as the horse, the origins of which can be traced through a clear progression of fossils, is more the exception than the rule. Even evolutionary certainties have their doubtful aspects. The dinosaurs certainly existed, but there is still much disagreement among paleontologists as to their behavior, diet, metabolism, and ecology. To put flesh on bones is not as easy as museum exhibits suggest.

As the Siskiyou forest demonstrates, though, living worlds are not always so different from worlds of long ago. They are sometimes surprisingly similar. If it seems incredible to many people that pre-historic algae and worms evolved into trees and people, it seems just as extraordinary to me that many algae and worms hardly have changed in 500 million years. Everything alive is a living fossil, in a sense; so it should be possible to perceive much of life's history by imaginatively projecting living landscapes onto primeval ones, much as the Greeks saw the mythical origins of their world in their living mountains and valleys.

The Greeks had a huge number of myths; each valley and island had its local pantheon. This is typical of the mythological way of thinking, and it holds true for evolution as myth. It's inaccurate to talk about a myth of evolution as though there were only one. There are as many myths of evolution as there are groups or individuals with differing responses to the scientific evidence. There is the establishment, mass-media myth, which presents evolution as a brisk upward sequence of floating, swimming, crawling, and walking shapes somehow leading inexorably to spaceships. There is the coevolution myth of the whole-earth counter-culture, which is just as progress-minded in its way as the establishment myth, but more democratic: it would have us take dolphins along on our spaceships. There is the recombinant DNA myth of the futurists, which would move evolution out of the biosphere and into the factory. Like the Titans from which the Olympian gods made the world, myth is huge and polymorphous. It can't be confined to the study of primitive and ancient societies. No sooner is a fact observed and recorded than it begins to be woven into myth's web of dream, imagination, and emotion. The bear tracks in the Siskiyous were facts that became myths.

My myth of evolution will be less orderly than some. It may be chaotic and devious, but this may be a more faithful reflection of reality than charts and graphs. Evolution, at least in the Klamath Mountains, is less a tidily consecutive array of increasingly advanced organisms than a leapfrogging mob of plants, animals, and dubious beings such as fungi, all earnestly photosynthesizing, feeding, respiring, and reproducing without much respect for hierarchy or direction. It is less a progression than a cyclic accretion wherein organisms appear or disappear for reasons that often are obscure or mysterious, and not readily applicable to scientific concepts. This is not to say that science is wrong, only that it is incomplete, as any scientist worth the name will agree. Whether a species as devious and chaotic as ours will ever achieve a complete science is in fact doubtful.

As told by its rocks and its living organisms, the story of life in the Klamath Mountains wanders in circles as often as it arrives anywhere, but considerable action and color tend to compensate for this lack of plot. The characters are unruly, prone to abrupt appearances and disappearances, and unwilling to submit to dramatic unities. Often they will not leave the stage after a climactic scene, but lurk upstage, sometimes making rude comments on the acts that follow. Their lingering makes the story more complicated to tell with each successive scene, since the stage continually gets more crowded. There are elements of epic and tragedy in this confusion, but the overall impression is more of comedy, even buffoonery, than high drama, more *A Midsummer Night's Dream* than *Hamlet*.

A story without a plot is not an easy one to tell. Like earlier myths, though, evolution can be divided into a series of ages or cycles. As classical mythology divided earth's history into ages of gold, silver, and iron, so evolutionists have divided it into ages of invertebrates, fish, amphibians, reptiles, and mammals. I'm not going to follow these conventional evolutionary ages, however, for two reasons. First, I think dividing evolution in conventional ages places too much emphasis on animals, which is understandable since that's what we are, but which obscures the fact that one-celled organisms, plants, and fungi are more fundamental to the evolutionary process. Animals, particularly vertebrates, are relative latecomers and are still vastly outnumbered by other organisms. Second, the conventional evolutionary ages don't take into account the fact that evolution is an accretion as much as a progression. It isn't as though invertebrates were no longer required after fish evolved. Invertebrates are probably more numerous and certainly more diverse now than they were during the age of invertebrates. They have continued to evolve, and many are just as "modern" as mammals, some more so.

If I'm going to approach the earth's history as embodied in something as hard to explore as the Klamath Mountains, I'll need more universal elements to symbolize the successive ages, elements as weighty and substantial as precious metals. So I'll use the most basic elements of the Klamath landscape to organize my untidy evolutionary story — rock, water, trees, and grass. Each has dominated the planet at some time: rock before life; water during life's several billion years of early development; trees since life emerged from water; grass since the planet began to get drier and colder, in the past thirty million or so years. Together, they still dominate the planet. Rock shapes the landscape; water erodes from rock the chemicals from which life is constructed; trees and grass make these chemicals into the foods upon which all life depends. Presumably, they will continue to dominate the planet if another, more recent element doesn't incinerate it; and I will have some things to say about this new, human element as well.

CHAPTER TWO

ROCK BOTTOM

. . . as if I were
Seeing rock for the first time. As if I were seeing through
the flame-lit surface into the real and bodily
And living rock. Nothing strange . . . I cannot
Tell you how strange: the silent passion, the deep nobility
and childlike loveliness: this fate going on
Outside our fates.

Robinson Jeffers
Oh, Lovely Rock

THERE NEVER SEEMS to have been any doubt that rocks came before living things — that they were, in a sense, the first beings. In the oldest myths rocks are tricky objects. Sometimes alive, or at least inhabited by spirits, they could move around and turn into other things. Monotheism quieted them down. They became rocks of ages, symbols of heavenly permanence and power, eminences for saints and prophets to stand on, foundations for temples and churches. Evolution seems to have reversed this trend toward quiescence, and rocks are on the move again. Although we no longer see them as animate, we know that some of them once *were* alive, that many will be alive again as their elements break down into soil and are taken up by plants, and that they are constantly on the move. Rocks have regained respect in the past century or two. They are not just inert stuff to be blasted through or piled up into buildings. They have a slow life of their own. They form, mature, and age, and their movements affect the quicker lives of plants and animals enormously.

Respected or not, rocks can't be ignored in the Klamath Mountains. Even in the deepest forest soil, one can't scratch the ground without hitting pebbles, and a lot of extraordinarily dense vegetation simply grows on rockpiles. Creek beds are rockpiles, slopes are rockpiles, valleys are stream-deposited rockpiles, crests and peaks are glorified rockpiles. The first time I backpacked into the Trinity Alps, I had to contemplate an 8,043-foot rockpile called Little Granite Peak for two days as I toiled uphill into the wilderness; and when I finally reached the Trinity crest, the peak seemed about as far above me as it had down beside Canyon Creek. I'd never had such strong feelings about a rockpile.

Cruel and mocking though they may seem to someone trying to get on top of them, the Klamath rocks are beautiful. They come in all colors — I was going to say "of the rainbow," but that's not true. There is little of the celestial in Klamath rocks: they have their own chthonic spectrum, which begins with the slightly soiled white of weathered marble, tones down to the silver of granite, passes through volcanic reds, serpentine greens, schist blues, and ends in the black of newly fractured peridotite, a midnight black that suggests the density of darkness that must prevail dozens of miles deep in the earth's mantle, where peridotite comes from. These colors are swirled and smeared across the mountains in a geological abstract expressionism that is as randomly intricate in microcosm as in macrocosm, so that a pebble has as many flecks and crystals of color visible to the eye as a distant peak. In a Trinity creek bed, with water rushing silver and amethyst and gold over the granite, it is easy to imagine an Ali Baba treasure cave and to understand the obsession with treasure that afflicted the miners who've swarmed the Klamaths since 1849.

Despite their beauty, I confess to an impatience with Klamath rocks. It isn't that they're hard to get on top of. I can take a mountaintop or leave it; the *depth* of a wilderness interests me far more than its height. It is that rocks are so much less orderly than organisms. I can't look at a pebble and call it a specific name that will identify it with a lot of other pebbles, as I can identify a California lady's slipper or black bear. I may call a pebble "feldspar" or "quartz," but there are so many kinds of feldspar and quartz that the names tell little about the origins and history of a particular pebble. A pebble in the Klamaths is likely to be made of feldspar *and* quartz, which means that the pebble has yet another name — granite — and the whole problem of identification begins again. And granite pebbles may have other minerals of obscure and equivocal origin in them. Of course, there are only so many atomic elements to go into a pebble, but their random combination won't provide much satisfaction to anyone short of a geophysicist.

Sometimes I wish the Klamaths were one of those placid areas where I could pass over the bedrock with a few learned remarks about inland seas or glacial action. Unfortunately, there's not much wilderness left on such placid terrain. There have been inland seas and glaciers where the Klamaths now stand, but there also have been most other geological phenomena known to science, and perhaps a few that are unknown. Klamath rocks don't sit idly under white sand as do Florida limestones, or roll gently under deep loam as do midwestern sandstones and shales. They are athletic rocks, at times prankish. They like to stand on their heads and play practical jokes, pranks unappreciated by a hiker who finds a trail ending in a landslide or a roadbuilder who sees a steel culvert tipping into a gully. As with all pranksters, it is hard to get a straight story from Klamath rocks; they prefer to speak paradoxes, obscure codes, or apparent nonsense. This is why geologists call them a "knot," a nightmare.

Rocks standing on their heads isn't just metaphor. There is hardly a mountaintop in the Klamaths that didn't start at the bottom of something. The gigantic white trapezoid called Marble Mountain began as a deposit of limy ooze under the Pacific. The red peridotite hulks of the Kalmiopsis were originally ocean floor and planetary mantle, beneath the limy ooze. The granite peaks of the Trinity Alps began deep in the continental crust. And what is at the bottom of the deep gorges that run below these peaks, so far below that peaks are invisible from gorges? In many places one finds twisted humps of black rock called pillow basalts which originated as lava flows from underwater volcanic action — molten lava setting into bulging pillow shapes on contact with seawater. This can seem very upside down to someone used to associating lava flows with the tops of volcanoes. Such strenuous flipflops (like old Father William standing on his head in Alice's poem recitation to the caterpillar) lead me to wonder how such an ancient planet can be so frisky.

Understanding the history of the Klamaths would be much simpler if the earth were a quiescent globe upon which successive geological strata placidly deposited themselves. One simply could dig through layers of sediment as archaeologists sift through prehistoric kitchen middens. There might be a few human skulls on top, like cherries on a cake, then fossils of flowering plants and mammals, conifers and reptiles, ferns and amphibians, algae and invertebrates, and, below them, the first lifeless sedimentary layers. One could know exactly which organisms inhabited a particular location for the five-billion-odd years since the earth's formation. But the more we know of earth's history, the less placid it appears. According to plate tectonics, the latest theory, the earth's surface is in constant, if ponderous, motion as at least eight enormous basaltic plates shift and jostle across it. Moved by convection currents in the earth's mantle the way slag is moved over molten steel, the plates grind sideways, collide, and sometimes ride up on one another like shifting ice floes.

The earth of plate tectonics seems more like the carapace of a turtle than like the tranquil blue globe of atlases and satellite photographs — a curious reminder of older myths in which a giant turtle supported the earth. It is a knobby, raggedy vision of the planetary surface, far from the ball-bearing elegance of Newtonian spheres. To make things even more irregular, plate tectonics sees continents as granitic smears riding on the basaltic plates like mud on a turtle's shell. And they don't ride quietly. As the plates have shifted, grown, and shrunk, the continents have been pushed together and pulled apart many times, and made to parade about the oceans like the floating islands of myth. There's nothing very subtle about this restlessness either. The blatant jigsaw-puzzle fit between Africa's west coast and South America's east coast that first made a few geographers suspect continents move was dismissed as a grotesque coincidence by most scientists until oceanographers discovered that the sea floor between the two continents gets progressively *younger* toward the center of the Atlantic: the continents are in fact moving apart as though on two conveyor belts.

Distressing as plate tectonics can be for those susceptible to vertigo, there is a chance that the theory makes sense of the Klamath knot. No one has made much sense of it before. Geologists knocked themselves out simply describing how Klamath rocks are distributed: in arc-shaped belts with the oldest rocks (Paleozoic era — at least 225 million years old) in the east, and the youngest rocks (Mesozoic era — at least 65 million years old) in the west. These arcs are made of sedimentary and metamorphic rocks, studded here and there with sheets of serpentine or peridotite from the ocean floor or granite from the continental crust. This description is a considerable achievement given the inaccessible and confusing Klamath terrain. The nongeologist sees no arc-shaped belts of rock, I can assure the reader. Still, the descriptions are really just a scientific way of saying that the Klamath Mountains are made of rocks standing on their heads, of formerly low-lying sedimentary, mantle, and crustal rocks that somehow have become peaks. They don't explain how or why.

Plate tectonics, though, gives even a geological simpleton like me a way of interpreting the Klamath knot. Geological maps of the region look to me like the results of a jammed conveyor belt. I once worked in an apple cannery where I noticed that the cans of apple juice got stacked up in just such arc-shaped belts when the conveyor belt was stuck (an interesting observation that was cut short when the foreman saw cans tumbling to the floor). The arcs of Klamath rocks, sedimentary rocks formed on the Pacific coast, may have been jammed against the western edge of North America by forces similar to those that splattered apple juice on the cannery floor.

If so, two crustal plates, one under North America and one under the Pacific, collided to give birth to the Klamath Mountains, which is almost as satisfyingly graphic as saying the mountains were thrown up by the conflict of two titans. The stronger North American plate, with its added weight of granitic continent, rode over the Pacific plate and forced its edge down into the earth's mantle, scraping layers of ocean floor sediment onto the continental margin in the process. Since the scraping process took millions of years (tectonic plates move at dizzying speeds of millimeters per year) and occurred in spurts, varying layers of sediment were scraped off, the oldest in the east, the youngest in the west. Also, as the Pacific plate was pushed down into the earth's mantle, it caused a severe case of geophysical indigestion, the sinking plate heating and deforming itself and the rocks it sank through as massive pressures and friction built up. Huge pieces of granitic continental crust melted, and were belched and vomited into rising mountains that had been ocean floor a few epochs before. Equally ponderous masses of the Pacific plate itself, serpentines and peridotites, were left behind in the mountains. The heat and pressure of the sinking plate also changed much of the new mountains' sedimentary rock into metamorphic schists, gneisses, and marbles (a process that complicated further the unraveling of the Klamath's geological history by destroying many of the fossils that might have determined the rocks' ages).

If this seems complicated, be assured that it is excessively simplified. It also is based as much on speculation as evidence. What is *known* about Klamath rock history from fossil and other evidence is even more tortuous. The mountains first began to form in the Jurassic period, which began about 190 million years ago. Geologists believe a vast continent called Pangaea was breaking up then, part of it moving west to become North America. As this part bumped into the Pacific crustal plate, the Klamaths may have been formed by the jammed conveyor-belt process. Actually, the Klamaths were just one part of a larger range that ran along the whole Pacific coast, including what are now the Sierra Nevada and the Blue Mountains of Oregon. Klamath rocks are very similar to Sierra rocks, as the forty-niners quickly discovered, finding the same gold-bearing gravels in both.

The mountain-building process begun in the Jurassic continued through the subsequent Cretaceous period, the Dinosaur Age. Then the westward movement of North America apparently dwindled, since sedimentary rocks stopped being scraped off the ocean floor for a time. The mountains began to be reduced by erosion. By the next period, the Paleogene, when dinosaurs had disappeared, the Klamaths were a region of broad valleys and gently rolling hills bordering a subtropical sea.

At this point some thirty million years ago, another period of tectonic activity began along the Pacific coast. This activity did not rejuvenate the eroded Klamath-Sierra range immediately. Instead, it divided it by warping the Klamath part westward and throwing up a vast volcanic plateau, the Modoc Plateau, between the Klamaths and Sierra. Rocks connecting the two ranges are believed to underlie the Modoc Plateau, but they are buried too deeply for geologists to be certain of their existence. It was not until fairly recently — two or three million years ago — that the Klamaths and Sierra became the impressive ranges we know today. A sideways rubbing of the North American and Pacific plates seems to have split them into large fault blocks rather as concrete pavement is split during an earthquake. Many of the blocks have been pushed upward into steep escarpments (the entire Sierra is one), and these formations have been carved by water and glaciers into the present's jagged peaks and deep canyons.

FOSSILS AND LICHENS

This rock history has had very little to say about the history of life. This is not because there was no life when the Klamath rocks were formed. The true Age of Rocks, called the Azoic (without life) era, came billions of years before any hint of the Klamath Mountains, or of what is now North America. The Azoic is so far away in time that not much of it is left on the earth's surface. In the over three billion years since life first appeared, Azoic rocks have been buried under miles of sedimentary rock almost everywhere. There aren't any visible within a thousand miles of the Klamath Mountains.

The Klamath rocks were formed when dinosaurs and primitive mammals were living, although there's no evidence dinosaurs and primitive mammals ever lived in the Klamaths. (There's no evidence they didn't, either.) There simply are no fossils of them. The meager fossil record that has been found in the Klamaths is largely irrelevant to the plants and animals there today. Sedimentary rocks are sprinkled with ammonites and other ancient marine invertebrates that have left few descendants anywhere on earth, much less in the Klamaths. An Oligocene epoch (about thirty million years ago) floodplain deposit near Weaverville on the Trinity River has yielded a fossil flora more suitable to the Florida Everglades than to contemporary California: leaves and branches of bald cypress, fig, and magnolia. Judging from fossils found in other parts of California, the Oligocene Klamath swamps probably supported alligators, tapirs, rhinoceroses, and many other present-day exotics (as well as a few still-native creatures such as squirrels and aplodontias).

When vertebrate fossils appear, in Pleistocene (less than two million years old) stream deposits near Douglas City, they are too late to reveal much about life's history in the Klamaths. Douglas City deer bones are not too different from bones of modern deer. Mammoth and ground sloth bones found with the deer add a certain fascination, since these shaggy monsters evidently browsed in willow thickets like those along the rivers today. But there is something a little incidental about the mammoths and ground sloths, as though they just happened to leave a few corpses behind during their glacial wanderings.

The recentness of the Douglas City bones reflects the recentness of the mountains. Today's Klamath peaks and gorges are not much older than *Homo sapiens,* for all their venerable quality. Their rocks are a great deal older, but even they probably aren't as ancient as the lichens that cover them in such variety of color and texture. (I once counted a dozen kinds of lichen on a single rock — black, gray, silver, green, orange, yellow — the same colors as Klamath rocks.) Algae and fungi are among the earliest organisms fossilized, so it seems likely that their symbiotic association in lichens is very old also, though I haven't seen any mention of fossil lichens. (Their stony habitat must be a barrier to fossilization; organisms need to be bedded down in soft sediments to fossilize well. These organisms live closer to rock than any others — they live by *devouring* rock with acids they secrete — but they're the least likely organisms to turn into rock.)

The durability of lichens is an example of the unusual evolutionary potential with which symbiosis, cooperation between different life forms, sometimes imbues organisms. Neither fungi nor algae grow well on bare, dry rock; it would have been self-defeating for them to compete for a habitat neither could dominate. Instead, chance favored an arrangement whereby they grew increasingly together, eventually forming "species" actually comprised of two organisms: a photosynthetic, food-producing alga and a fungus which holds the algal cells in a spongy matrix of mycelial threads. There is still some doubt as to whether the algal part of a lichen really benefits from the fungal part, whether it is not simply an "unwilling captive" of a fungal parasite, but this seems a semantic quibble. The abundance of lichens on sun-blasted, frost-cracked Klamath rocks attests to the well-being of the "captives."

Lichens point up a revolution of human consciousness that evolution has caused — the realization that life forms are much more durable than land forms. A species of lichen is more of a monument than the peak it encrusts. Tectonic theories of drifting continents and colliding crustal plates have completed the undercutting of traditional notions of rocky permanence, an undercutting begun by Darwin's precursors in the eighteenth century. Rock may have come before life, but it is a shifting foundation, at the bottom of the living world only in the limited sense that it is at the bottom of land and sea. Life has been around so long that its remains often lie deep below what we call "bedrock." The sedimentary rocks of the Klamath Mountains are estimated to be ten miles deep, so conceivably there could be fossils of Paleozoic creatures buried ten miles deep in them. That would seem a suitably permanent grave, but erosion probably will reach and disintegrate even those fossils someday, and then lichens will grow on them.

Today's Klamaths are very little like the frosting on a layer cake, then. They are more like an entire, heavy cake that has been dropped on a previous one, splattering it all over the table. By their very formation the mountains have obliterated most traces of any landscapes that preceded them. Even if the Klamaths were loaded with recoverable fossils, we probably couldn't learn much about their present life from those fossils. Life forms may be more durable than rock, but they also are even more restless, and the fossilized ancestors of today's Klamath flora and fauna must be scattered around the globe. Bears, for example, appear to have evolved in India.

Evolution may have undercut the bedrock foundation of human tradition so completely as to stand it on its head, like old Father William. Instead of rock being the standard of permanence, evolution seems to decree that only soft, yielding things shall persist on the earth. It's doubtful Jesus had lichens in mind when he said the meek would inherit the earth, but his prophecy makes good evolutionary sense. Mountains pop in and out of the planet like acne on a nervous adolescent, but the same kinds of lichens grow on their rocks. If this is indeed a general rule of evolution, then the increasing resemblance of human settlements to mountain ranges might give us pause.

Instead of striving so adamantly to rise above the competition in towers of steel and stone, civilization might endure better using the lichen model. Lichens flourish in hard places by close interpenetration of food-producers and food-consumers. The fungus parts of civilization, the institutions and governments that control and protect, too often lose contact with the alga parts, the food-producing rural communities. It is easy to forget, from a tower, that civilization is based on the symbiosis that is agriculture. We did not "invent" farming as we consciously develop technologies today; it evolved from interactions between Neolithic communities and the wild plants and animals they lived on.

The Klamath Mountains have felt the force of civilization's obsession with exalting on rockpiles in the past century. Precious metals from their creek beds helped raise the towers around San Francisco Bay and the Los Angeles Basin. Miners ransacked the Klamaths as violently as they did the Mother Lode in the Sierra. I'm in awe of miners' work, more than that of glaciers and volcanoes; it's harder to conceive of human hands leveling hills or scouring out stream beds. There's something titanic about the way miners fasten on a piece of earth and chew it up. I remember a little, swarthy gold miner I met in British Columbia. With his little, blonde wife and his mushroom white hardhat, he might have been one of the dwarfs from *Das Rheingold*, though he lived in a trailer, not a cave. His life was epic. He showed me a black bear track and told me it was a lynx's — the lynxes of his mind were bigger than he was, and he never stepped outside without a 30.06 and a bowie knife. The next claim over was run by a group of Texans so expensively equipped, with uniformly white and baby blue hardhats, overalls, and placer machinery, that they might have been an interplanetary exploration team.

Miners I've met in the Klamaths have been less colorful, but equally determined. A middle-aged couple who were staking a claim halfway up Virgin Creek in the Trinitys carried enough rifles and pistols to repel every bear in the county. (They complained of not seeing any wildlife.) A young couple had been living four years in a cabin on the New River that was reachable in winter only by sliding across the river on a cable. This present activity is only a faint echo of the mining that went on once. Many river beds are now undulating heaps of cobbles left by big companies that washed away entire bluffs with giant hoses and extracted millions of dollars' worth of gold.

I'm amazed how far into the wilderness the old miners got. After hiking five days into the Trinity Alps, the largest wilderness in the Klamaths, I came upon a stone dam built by a mining company to supply water for hydraulic mining several dozen miles south on the Trinity River. Some of these old mining sites are eerie, like Taggart's Bar in the Kalmiopsis. Taggart's Bar is on the Chetko River, which runs between steep cliffs of volcanic stone and is only accessible from above in a few places. I camped several days on the cliffs and have rarely felt such wildness and loneliness in a place. Overhung with mossy oaks, the gorge has an enchanted quality, as though there might be dragons' dens under the cliff. A small bear almost walked into me on the path, so little did he expect to find people there.

But there once was a town on the cramped, shadowed flat beside the river; I stumbled one evening on stone foundations so covered with fern and alder that, unseeing, I'd walked past them many times. Historians think many of the early miners who worked the Chetko were Chinese, but the stones told nothing about the people who'd lived there. Considering its remoteness even today, at least a long day's fast walk from a trailhead, it must have been a strange community, perhaps like the ones Ambrose Bierce described in his gruesome tales about the Mother Lode. There is an old legend in the Klamaths that a mining town on the Chetko was attacked and destroyed by shaggy giants that swarmed out of the hills, enraged because a miner had killed one of them.

The early miners were so ubiquitous that a weekend prospector I met on the Trinity River said he found most of his gold attached to relics of old mines. Gold dust coats clots of mercury and rusty nails that have been lying on the river bottom since escaping from some turn-of-the-century placer rig. The relics are not all so old. The Kalmiopsis contains fully equipped chromium mines left over from World War II that look as though they could be started right up again with a little grease and rust remover. One of the prettiest little marsh-meadows I found in the Kalmiopsis — a botanical jewelbox of leopard lilies, cobra plants, azaleas, white bog laurels, sneezeweeds, and coneflowers — was perched between an old bulldozer scar and a rusty but intact steam shovel of a kind I hadn't seen since I was a child in the 1940s.

An iron behemoth overrun by wildflowers shows the vanity of human attempts to attain what we've fancied to be the strength and stability of the mineral world. The flowers will be there long after the steam shovel has rusted into thin air. At the same time the steam shovel demonstrates the weakness of rock when confronted with soft human brains. Nobody has yet devised a way of grinding up entire mountain ranges for mining purposes, but with gold nearing a thousand dollars an ounce as I write this, who knows? We are leveling Appalachian hills to get the coal under them.

Life has achieved a certain independence from rock. Though its original elements came from the planet's rocky core, the physical matrix in which life now exists — atmosphere, ocean, soil — is largely a product of metabolism. The coal and petroleum that fuel our mountainous civilization did not originate as rock (despite the name "petro-oleum" — rock-oil) but as fossilized swamp plants and plankton. This is not to say that geological movements don't affect organic evolution. The creeping of crustal plates can turn a lush continent into a vast glacier such as Antarctica (where discovery of a fossil reptile already known to be from Africa was an early proof of continental drift).

Despite its great importance as a preserver of fossils, rock alone can't elucidate the planet's history. Plastic and restless, it conceals or destroys more fossils than it preserves. The fossil record also is too generalized to apply to particular places with consistency, and particular places is what the living world is made of. Trying to see the ancestry of the Klamath Mountain ecosystem only through rocks would be like trying to understand a human life from only bones. Much can be learned from bones, but it is preferable to get to know people while they are still alive.

Of course, there is always doubt in relating what is to what was; but if living things lack the solidity and fixedness of fossils, they at least share the basic problem of survival with their ancestors. They wouldn't be here today if their ancestors hadn't solved it. Organisms outlast mountain ranges because they work much harder at existing than rocks do.

CHAPTER THREE

PRIMAL OOZE

It is not far from here, measured in miles, that the mere stands;
over it hang frost-covered woods, trees fast of root close over the water.
There each night may be seen fire on the flood, a fearful wonder.

Beowulf

IF EVOLUTION RESEMBLES animist mythologies in its view of rocks, it
departs from them in the matter of life's origin. Almost all early
mythologies saw mud or dust as the basic stuff of life, which is sensible, since dead things turn back into mud or dust when they decay,
and since mud seems to breed new life in the form of flies and
worms. Evolution, though, sees water as the source of life, as both
the medium in which it first formed and its continuing major con-
stituent. The idea, which chemistry has verified, that even bone and
muscle are mostly water would have seemed bizarre to a Sumerian.
Ancient worlds were surrounded by oceans, but the idea that life
began microscopically in the water and then gradually moved to
land was not current before evolution. Water was certainly a primal
element in ancient myth, as when Jehovah moved across the face of
the waters, but it didn't generate life directly. An outside force was
required to impose life's solidity on it. Water was feared as an alien
abode of untrustworthy spirits as well as revered as a life-giving
element.

The mythic strangeness of water echoes over the lakes of the Klamath Mountains: over Spirit Lake, Man-Eaten Lake, Secret Lake, Lost Lake, Deadman Lake, and Wild Lake in the Marbles; Knownothing Lake in the Trinitys; Devil's Punchbowl in the Siskiyous; Lonesome Lake in the Red Buttes. Mountain lakes have an old reputation as abodes of strange, often monstrous, beings. Indian myths of the Klamath region mention giant serpents with human faces that live in lakes and prey on unwary land creatures. Giant, humanoid tracks like those found near Bluff Creek in the Siskiyous have been reported around mountain lakes. Sometimes the tracks lead to holes in the ice, but not away again.

There are deep places in Klamath Mountain lakes where even the intense California sunlight shows no bottom. The forest around Spirit Lake is pocked with sinkholes, where marble bedrock has subsided, eroded by underground waters. Deep caverns could underlie such a lake. A few dozen air miles to the north of the Marble Mountains is Oregon Caves National Monument, where one walks past stalactites and stalagmites to the verge of underground waters so utterly black that a flashlight beam seems not to penetrate them, but to be absorbed.

Something in the human mind resonates at the image of a dark, still, conceivably fathomless mountain lake. The fascination isn't confined to ancient myth. A scientist with whom I was discussing the Siskiyous was intrigued by reports of unidentified, giant salamanders in high Siskiyou lakes. In our modern insulation from the natural world, we don't fear the fascination of mountain lakes as our ancestors did, but the basic impulse is similar, as though we carry a molecular memory of our kinship with water life, a memory that would have seemed strange and fearsome to people unfamiliar with evolutionary concepts. Mythical visions of giant serpents in bottomless lakes are not much different from the dreams I had, and these dreams could as easily have been frightening as reassuring.

One scientific theory says life began in mountain lakes. Lakes in Azoic era mountain ranges would have been closer than the seas to the lightning and ultraviolet rays that presumably caused the complex hydrocarbon and protein molecules of primal ooze to form. The rich, proto-organic chemicals that comprised the first hydrocarbons might have been scarce in the bare rocks of primal mountains, though, and the lake theory of creation is not popular today.

It seems more likely that life began in warm, shallow seas, invading inland waters much later.

Klamath mountain lakes make this unpopular theory seem plausible, however — so much life comes from them. To stand beside one on a hot summer day is to witness a fecundity incongruous with the austere pine and fir forest around it. Whiffs of methane bubbling from the gray green bottom taint the piney air. Dragonflies and damselflies weave spells across a surface dimpled with trout strikes and littered with dead insects of a hundred kinds. Shallows are choked with sedge and festooned with algae; deeper water is packed with pondweeds and stoneworts; and grasslike spikes of quillwort punctuate the ooze well out into deep water. Even the mud and pebbles of the shore are crusted with a viridian slime of blue green algae. It seems that life could be created spontaneously from such ferment, hydrocarbons and proteins woven from ooze into brand new organisms. But life itself evidently closed off the option of new creation billions of years ago by generating the oxygen atmosphere — a powerful poison to the first anaerobic organisms — and by screening out most ultraviolet radiation with the ozone layer.

Even so, a mountain lake's air of the primordial is not simply an illusion. Its fresh, acid waters may be more like the primal oceans' than today's Pacific. Some inhabitants of Klamath Mountain lakes might have been at home in waters of 4 billion years ago. The oxygen atmosphere may have stopped the creation of new life, but it has not stopped the first way of life. Deep in lake ooze, where decay has exhausted the oxygen, anaerobic bacteria continue a prephotosynthetic existence so ancient it probably preceded the earliest known sedimentary rocks, which are in Greenland and are thought to be 3.8 billion years old. The earliest known fossils, 3.4 billion years old, are of fully evolved anaerobic cells, so life must have appeared well before that date. It may have appeared soon after the earth became capable of supporting life.

The methane marsh gas that bubbles from today's lake ooze was a major component of the primordial atmosphere. I've always been fascinated by will-o'-the-wisps, luminous clouds of methane sometimes hovering over boggy places on dark nights. It's appropriate that this ghostly phenomenon should be the exhalation of our most remote ancestors from their burial place in the airless ooze. I've never seen a will-o'-the-wisp in the Klamaths, but there's no lack of methane. A foot placed into the ooze brings up great wallowing bubbles redolent of dead fish and rotting algae. The ooze is warmer than the lakewater, a graphic indication of the biotic activity it harbors.

Some old Klamath lakes are so filled with ooze one can walk across them. I crossed one called Azalea Lake in the Red Buttes and never touched a firm bottom, only the green ooze into which my feet sank until the water's buoyancy neutralized my weight. If I'd been heavier, I suppose I simply would have sunk deeper, who knows how deep or past what buried objects, serpent ribs or giant skulls. The lake was deceptive; its clear water made it seem shallower than it was, and the ranks of quillworts (a primitive water plant related to fern) made the bottom look firm. Buried tree branches pricked my feet. I felt as though I might be standing atop an entire forest sunk in ooze.

Azalea Lake exhaled eerie morning mists. Vapor rose from it in little wisps before sunrise, then the wisps coalesced and rolled upward in sheets as the first solar rays touched the treetops. By the time the sunlight reached the lake surface, the mist became so thick that the surrounding forest disappeared, leaving only water, mist, light — the first living landscape.

It is not really the first landscape, of course. The mists of that time billowed in methane, not air as we know it. Still, the oxygen-rich air above Klamath lakes is a heritage of early organisms just as devoted to ooze, and just as common in it today, as anaerobic bacteria. Blue green algae live very quietly in Klamath lakes, becoming noticeable only as disagreeable-looking crusts and mats in warm, shallow water; but ancestors very like them probably were the first photosynthetic organisms, and thus created the oxygen atmosphere. Disagreeable-looking crusts and mats perhaps made by blue green algae have been found fossilized in African rocks almost three billion years old. Blue green algae definitely are much more primitive than other algae, since they lack cell nuclei and chloroplasts (organelles which package green chlorophyll in all other photosynthetic cells, from algae to oaks). They resemble bacteria, which have been found fossilized with them in extremely ancient rocks. Blue green algae and bacteria probably comprised the first ecosystem, with algae as food producers and bacteria as predators, scavengers, or parasites.

This ecosystem is still working in Klamath lakes, where bacteria may number a million cells per liter of warm water. Other very primitive organisms called actinomycetes, which resemble the branching mycelial cells of fungi except that they too lack cell nuclei, may number ten million cells per liter of ooze. Primitive though they are, bacteria, blue green algae, and actinomycetes are not exactly shadows of former grandeur: they proceed with their simple lives as though evolution had ended as well as begun with them. They outnumber everything else.

Actually, there is not certain proof that evolution *did* begin with them. So far there are no fossils linking them with higher life forms. When higher life forms appear, life is so far along that it might have arrived from a falling star. Australian cherts of 900 million years ago contain fossils of true algae and fungi, and Australian sandstones of 800 million years ago contain evidence of jellyfish and flatworms. It seems likely that the ancestors of bacteria and blue green algae evolved into the cells with nuclei that comprise algae, fungi, worms, and jellyfish. Bacteria move around like animals; blue green algae are photosynthetic like plants; actinomycetes are like fungi (which really are very different from animals and plants). It's not easy to see *how* this evolution took place, though.

How did cells with something as complicated as the nucleus evolve from cells without it? The nucleus gives the cell a great evolutionary potential that bacteria and blue green algae don't have, as evidenced by the fact that all higher organisms are comprised of cells with nuclei. How did that potential arise? Gradual changes caused by mutation and natural selection of DNA molecules hardly seem to account for such a leap, no matter how many billions of generations of blue green algae and bacteria reproduced and died in the primal ooze.

The appearance of the nucleus may have been one of the first times that life changed not through death, through naturally selective culling of unfit organisms, but through symbiosis. There are no fossil links between cells with and without nuclei, but there is a *living* link of sorts, one so common that it lives in every lake and puddle in the Klamaths. A magnified drop from one of these will contain spindle-shaped creatures that swim about propelled by whiplike tails, but that also are green with photosynthetic chlorophyll. They are both plant *and* animal so far as biologists can tell.

Their composite nature suggests that, as well as competing among themselves, various early bacteria and blue green algae evolved cooperative ways of life. The cell with nucleus may have originated as a composite of several different kinds of cells without nuclei, somewhat as the many-celled organism is a composite of many different kinds of cells with nuclei.

Some ancient, giant bacterium that went around eating blue green algae and smaller bacteria may have "discovered" that if it did not digest its prey immediately but allowed it to go on living and producing food or other metabolic byproducts *inside* it, it would have an easier life than before. (This might be compared with the first domestication of plants and animals by humans.) Gradually these "captives" would have lost their individuality and evolved into the nuclei, chloroplasts, mitochrondria, and other internal structures of the cell with nucleus. Chloroplasts and mitochondria still reproduce apart from the cell nucleus, as though retaining one feature of independent origin.

If symbiosis is so fundamental to life as it exists today, I wonder what it will do to life in the future. Will present symbioses eventually become super-organisms, the individual members of the relationship melting into one another? Those chicken farms where thousands of birds live in huge rooms, their feet never touching the ground, raise bizarre thoughts along these lines. Highly technologized human civilization is *absorbing* organisms; such chickens have hardly more independent existence than a chloroplast's. We look upon such relationships as gruesome and unnatural — chickens treated as machines — but perhaps there is creative potential here, if we learn to keep our absorptive tendencies in balance with the biosphere and don't try to swallow the world. World swallowers fare poorly in all myths. The gods tear them apart and make new worlds with them, which is more or less what happened to that primevally large and greedy bacterium. It set out to get rich by making captive blue green algae and weaker bacteria work for it, but ended working as a galley slave in the fleets of worms, jellyfish, and, much later, industrial chickens.

Lack of fossil evidence makes all this pretty unscientific, of course. There is not even fossil evidence that many-celled animals evolved from single-celled organisms. Some single-celled organisms are more highly evolved than some many-celled organisms. Fossils indicate that the gemlike, single-celled diatoms evolved long after the many-celled green algae. We assume that single-celled organisms evolved into many-celled because it seems reasonable, but if one strips away the reasoning, evolution is not so unlike older myths in describing life's origins. All contain a good deal of murk and bluster, and are peopled with mysterious and amorphous beings. The fact that evolution's primal denizens — its titans, so to speak — are too tiny to be seen with the naked eye does not make them less impressive. Despite its reduction in some minds to a plaything for schoolchildren, the world of single-celled organisms is formidable, at once too distant for understanding and too close for comfort.

Time and place do not exist for the microscopic world in the same way they do for the many-celled. Microbes inhabit microclimates, which have a different geography and chronology than our climates. The same microclimates may exist in a tundra puddle and a jungle treehole, and the same microbes will live therein. And microclimates don't exist only in lakes and puddles, they exist within many-celled organisms as well. The invisible titans haven't remained in primal ooze, they are in us, they *are* us. The eye that looks through the microscope teems with more cellular life than the water drop on the slide.

It is only a convenient fantasy to look for the first organisms in the lake ooze. Bacteria cover every surface, tree root and bear tooth. Green algae bloom on Klamath Mountain trails after every rain and somehow contrive to live and presumably carry on photosynthesis, several feet underground. One could invent a biology in which many-celled organisms did not exist, only different kinds of microbes, colonial or independent. But the landscapes of such a world would be indescribably alien to many-celled human senses.

SWIMMERS

Remote descendants of the first many-celled animals are common in Klamath Mountain waters. In the Marble Mountains, in Meteor Lake, I found a twiglike object of such a bright emerald green I thought it a St. Patrick's Day swizzle stick. Looking closer, I saw it was a twig coated with green jelly in which tiny, glassy spikes were embedded. This was a freshwater sponge — the jelly, its algae-riddled body; the spikes, its silica skeleton. We're not sure if sponges are the oldest many-celled animals, but they are the simplest: mere cell colonies without nervous systems or muscles. Next door, in Monument Lake, I looked through a magnifying glass and found a pinkish spindle shape of clear jelly with tentacles that grew longer as I watched, until they were several times the creature's body length. It was a hydra, a miniscule relative of the great ocean jellyfish.

I'd have dismissed tiny lumps of dusky tissue I found when idly turning over rocks in rivulets as "insect eggs" if I hadn't had a magnifying glass. I discovered they were flatworms, relatives of the droll planarians that come after hydras in textbooks. The biology lab associations of flatworms made it seem strange to find them in a wilderness, even though flatworms have inhabited wilderness at least 800 million years and biology labs for little more than a century. I'd seen wild flatworms before, in Ohio swamps, but they had been grayish and arrow-shaped like ordinary textbook flatworms. These Klamath flatworms were bizarre, with tiny heads and fat, speckled bodies. There were corpulent pink ones and thinner green ones, and the green ones seemed to follow the pink ones, like boars after brood sows. The cold mountain waters were a long way from warm seas of 800 million years ago, but the flatworms obviously were bursting with health, their flanks glistening prismatically in the intense sunlight of six thousand feet above sea level.

The worm, lowliest animal in the medieval great chain of being (a position still reflected in our everyday speech), plays a venerable role in evolution. It seems to have been the first organism to fully behave like an animal: with efficiently coordinated locomotion, digestion, circulation, and respiration. Its role in ecosystems is almost as basic as those of bacteria and algae. If all other life vanished from the earth, transparent roundworms that live in soil and water and parasitize most other organisms would still give ghostly shape to the biosphere. Annelid worms, more evolved than flatworms and roundworms, include not only earthworms but even more abundant aquatic worms such as tubificids, which live in cold, stagnant lake bottoms, waving their tails in the water to circulate food and oxygen. Because few other organisms can tolerate the abysmal conditions of deep lakes, tubificids may number eight thousand per square meter. Deep-water worms such as these, which respire in oxygen-poor surroundings, may have been the first to evolve red, hemoglobin-rich blood.

The ancient Chinese could have had worms in mind instead of reptiles when they invoked ancestral dragon spirits; the bristly sand-worms of coastal ooze look more like dragons than do snakes or lizards. Each of the three great animal groups dominating the world today — mollusks, arthropods, and chordates — displays traits of worm ancestry. Not only blood, but muscles, nerves, sense organs, and gills are shared by worms and these groups. Primitive species and fetal stages of all three groups resemble worms. Many were thought to *be* worms before evolutionary classification sorted them out.

The worm origin of all three groups is a blow to the myth of evo-lution-as-hierarchy, the modern great chain of being which we have unconsciously read into evolution. If each group evolved independ-ently from worms, it makes no sense to put vertebrates on top of mollusks and arthropods. Fossils demonstrate no such progression. On the contrary, vertebrates such as fish and amphibians seem from fossils to have appeared before some arthropods, including insects and spiders. In a chronological sense, at least, they are thus more primitive. The picture the progress-minded still like to paint — life crawling from the water, reaching toward the stars — is a caricature of evolution. Life has been jumping back and forth between water and land since passing the worm stage, and will continue to do so.

The three major phyla live in biotic democracy, and extraordi-nary abundance, in Klamath Mountain waters. In a little tarn called Rock Lake at about six thousand feet on the Salmon-Trinity crest, I found the tiniest mollusk I've seen, a pinhead-size pill clam with a golden shell. As I watched through my magnifying glass, it cautiously pushed its transparent foot between its two shells and crept along the bottom of a metal dish looking for some ooze to burrow into, perhaps a patch of silt eroded from one of the 100-million-year-old clam fossils common in some Klamath rocks.

Sharing the lake with the clam was a cross section of arthropods that closely resemble their Paleozoic relatives. What looked like a greenish, jet-propelled clam was an ostracod, a shrimplike crusta-cean that somehow has evolved a bivalved shell much like a clam's, but rows about with its legs like other crustaceans. Fossilized in rocks 500 million years old, ostracods remain so common that I got dizzy watching dozens revolve like miniscule sputniks. Even more numer-ous were stout, golden water fleas — vaguely teardrop-shaped, big-eyed crustaceans that swim on their sides, propelled by branching antennae which they hold opposite their heads as though ready to punch themselves in the nose. They swarmed around the sedges of the lake margin. Burrowing in the lake bottom were greenish amphipods, resembling ocean shrimp, which burst up like quail from a thicket each time I took a step, then hastily burrowed out of sight again. When I walked a few yards toward the lake's center, bright red copepods ballooned to the surface on methane bubbles I kicked up. Torpedo-shaped crustaceans with man-from-Mars antennae, copepods may number a thousand per liter of lake water.

Every lake I've examined in the Klamaths has been some such alphabet soup of tiny creatures, and their prevalence is especially impressive when you consider that the lakes are little more than ten thousand years old, having been formed by the last glaciation, while the mollusks and crustaceans have lineages about fifty thousand times that. How do such small, seemingly passive creatures so abundantly populate such recent waters? Many probably arrived on the wind as eggs or cysts. Others may have arrived attached to more mobile organisms, such as fish. Rainbow trout had entered many Klamath lakes by their outlet streams before we started stocking them by airplane. At Meteor Lake I found good-size rainbows trapped in pools of a drying outlet in late summer, casualties of some seasonal urge to leave or return to the lake.

The poor trout smothering in their muddy backwater (I tried to rescue some but they were slippery and stubbornly evasive, actually jumping on land to avoid capture) seem emblematic of the transition that separates the oldest, purely aquatic members of the three major phyla from more recent ones that spend at least part of their lives out of water. Clams, crustaceans, and trout may have traveled far to populate Klamath lakes, but not as far, figuratively, as other creatures. Life's move to land hasn't been quite as portentous an event as certain land-dwelling minds would have it, but it's important enough that it has never stopped occurring.

The most conspicuous ooze dwellers of Klamath lakes are not purely aquatic, but amphibious. They inhabit both water and land because they never emerged completely from water, or because they emerged completely from it once, then went back into it. The first category is ecologically the less important of the two, but it has a strong fascination for us because it includes the first land-dwelling vertebrates, which we presumptuously call "the amphibians" as though no other creatures but vertebrates divided their time between water and land.

Fossils indicate that amphibians evolved from the lobe-finned fish during the Devonian period, as much as 395 million years ago. Lobe-finned fish had lungs and four ventral fins resembling flippers, the air-breathing lungs having evolved from swim bladders which serve most fish as flotation and pressure-regulation devices, and the flipperlike fins from the ventral fins most fish have. These lungs and fins were adaptations to life in the shallow seas, lakes, and swamps of the Devonian. The lungs allowed the lobe-finned fish to get extra oxygen in stagnant water by rising to the surface to breathe, while the "flippers" allowed them to move around on shallow bottoms more effectively than other fish.

The story of the lobe-finned fish says something very important about evolution. By evolving lobe fins and lungs, these ancient fish were *not* evolving toward life on land. Evolution is not directed. There was no imperative call 395 million years ago for fish to progress to land. Lobe fins and lungs were adaptations to life in *water,* not on land, and very efficient aquatic adaptations they were. Lobe-finned fish remained abundant for hundreds of millions of years, and persist today in the coelacanths of the Indian Ocean. The lobe-finned fish's move to land was an accident, doubtless an accident that occurred over millions of years, but still an accident. They were not *adapted* to life on land. How could they be, always having lived in water? They were *preadapted.*

Natural selection is propelled by billions of tiny accidents, depending on which individual happens to get which combination of genes. Preadaptation consists of gigantic single accidents, evolutionary jackpots. It occurs when an organism happens to have some trait, evolved for survival in its ancestral environment, that suddenly gives it an advantage when its environment changes. Environmental change means death and destruction for most species; only a tiny few are ever preadapted. But these few have had a disproportionate influence on the planet. Terrestrial life as we know it wouldn't exist if lobe-finned fish hadn't been preadapted. Civilization wouldn't exist if humans hadn't inherited binocular vision and opposable thumbs from simian ancestors. Adaptations to arboreal forest life, to grasping branches and gauging leaps, these traits were preadaptations to tool use, to hefting clubs and throwing rocks, and by extension to written language and technology.

At some point lobe-finned fish found themselves in the same predicament as the pathetic trout at Meteor Lake. They were trapped in drying swamps and lakes during a period of increasingly arid climate. Since they could already breathe air and walk on lake bottoms, though, they didn't die, but instead began walking overland in search of pools not dry. As the land wandering continued over millennia, legs and other amphibian features would have evolved through natural selection, completing the evolutionary leap.

The first amphibians were scaly and snaggle-toothed, like fish. In some ways modern amphibians, frogs and salamanders, are a greater departure from their first ancestors than scaly, snaggle-toothed reptiles are. Frogs and salamanders are better adapted to mountain lakes than their ancestors were — to backwaters without tide or current, poor in oxygen, and doomed to be choked with silt in a geological instant. They can breathe through their skins as well as with lungs and perform surprisingly precise overland journeys. In a sense they are better adapted to lakes than even purely aquatic creatures, and they frequent them in mobs that some people find repugnant or ominous but others find a source of peculiar fascination.

Hardly a lake in the Klamaths is without legions of rough-skinned newts — six-inch, lunk-headed salamanders with orange or yellow bellies. Their other name, water dog, attests to this abundance, since uncommon things seldom are called dogs. Newts spend spring and summer in lakes, where they breed, then take to land and spend winters underground. Rainy nights in spring and fall are times of mass newt migration during which one may stumble over a newt every few yards along a trail. A migrating newt is awesome, a rubber toy of a creature waving spindly legs as though it's a clockwork toy, but persistently climbing almost vertical slopes through a tangle of leaf litter and herbage that soon would exhaust a man of comparable size, its silver-flecked eyes unfocused on any conscious goal or reward. Spring-migrating newts seem to be moving away from the nearest body of water as often as toward it, but their numbers in summer lakes belie this impression of random movement. Experiments indicate that newts return to the water in which they hatched, and that they often return even though blinded, mutilated, or removed some distance.

I've seen the prune-size egg cases of newts attached to horsetail shoots, which is appropriate since horsetails are plants virtually unchanged from ancestors of Devonian swamps. Newts are classed among the more primitive salamanders because, like ancestral amphibians, they have lungs and lay their eggs in water. If lung-breathing and water-breeding seem contradictory as a way of life, this is complemented by the biology of more advanced salamanders, which have lost their lungs, but live and breed on land, respiring through their skin. Evolution doesn't have to be consistent, it merely has to work.

It's evident newts have lungs because they rise to the surface periodically to swallow air, supplementing the poor oxygen supply they can absorb through their skin on lake bottoms. I spent an afternoon at Babyfoot Lake in the Kalmiopsis watching newts rise for air in water clear enough to see them perfectly at a depth of ten feet. For a newt to rise ten feet takes a while, and I became semihypnotized as I lay on a floating log, watching a tiny newt face grow larger as its owner undulated upward. Indian myths say one can be hexed by a newt's eye, also a popular ingredient in medieval witches' brews. The spell was broken when the flat head and expressionless eyes broke the surface, and the newt took a gulp of air with the sound of a small bubble breaking, then started back toward the bottom.

As the afternoon drifted on, I began to feel pleasurably newtlike, lounging on the warm gray log with limbs dangling and head empty. I took to paddling along the surface, as some newts were doing (I watched one thrust its snout into a floating clot of fir pollen as though to inhale the bouquet), but when I tried to dive down and join the majority on the bottom, I found the water unbearably cold, as though the vast, chilly hand of newt antiquity were squeezing my mammalian chest. Newts do not have to breathe by expanding and contracting their rib cages like other vertebrates, because they excrete carbon dioxide through the skin.

A marshy pond called Lilypad Lake in the Red Buttes held the densest newt population I've seen. It lies under the south face of Kangaroo Mountain, a windy, dewy place of steep meadows and red fir groves, with a hint of melting snow in its air even in late summer when snow is long gone. Half the Klamath Mountains are visible from it — the Marbles, Trinitys, and high Siskiyous — and the cone of Mount Shasta floats in the southeast. Lilypad Lake seems incongruous with all this grandeur: smothered in yellow water lilies, it seems more like a Monet pasture pond than a wilderness lake, an impression enhanced by the cowbells and smells of summer-pasturing cattle. There was about one newt per square foot, or one newt per lilypad.

Brownish tadpoles of the red-legged frog shared Lilypad Lake with the newts. Frogs are even better adapted to lakes than newts, as is evidenced by their nearly worldwide distribution. Red-legged frogs plunge into the iciest lakes so readily at one's approach that the plop of their dive is usually the only betrayal of their presence. After spending an infernal July morning climbing the Sawtooth Ridge in the Trinity Alps, I was surprised to find the heather bogs around Caribou Lake still largely snow covered, and even more surprised to find red-legged frogs in full cry in the bogs. They carried on all day, a din that seemed unnecessarily assertive, like the hearty bellowing in snow outside a sauna. Sweaty from the climb, I tried to bathe in a bog pool, but hopped out almost as fast as the frogs had hopped in. The iciness gave me a new respect for organisms capable of carrying on in such surroundings.

For all their mating racket, frogs are furtive about emerging from their underwater tadpole stage, as though not fully convinced of the wisdom of the transition. Little frogs lurk on shore in scattered numbers, always ready to hop back underwater. There is no such timidity about the western toad, a creature whose adult independence of water for everything except mating is complete.

Young toads make a huge spectacle out of emerging from the water. On exactly the last day of summer, I saw one such emergence at already-frost-touched Little Elk Lake in the Marble Mountains. I'd planned to camp at the lake, but the first thing I saw on arrival was a large bear and her cub. A dozen deer hunters already camped there said she'd been trying to raid their camp, and if that weren't discouraging enough, they added that the lake would be a "war zone" when deer season opened the next morning. Departing, I crossed the lake's outlet and suddenly noticed that most of what had appeared to be running water was a huge, glistening mass of tiny, pot-bellied toads shuffling nose-to-tail out of the lake like rush-hour traffic. I could have shoveled barrelfuls of them.

I walked to the lakeshore and found it a continuous toad beachhead, the toads walking out over a mat of dead and dying black tadpoles. A solid mass of tadpoles thronged the shallows, as though impatient to become toads in their turn and anxious to be out before the winter ice. When I picked up a toad, it kept right on pumping its legs, undaunted in its determination to get somewhere. There were hundreds of thousands of toads in that lake.

Superabundant as they are, tadpoles are an evolutionary mystery. Newly hatched Paleozoic amphibians were small replicas of their parents except that, like their fish ancestors, they had gills. Somehow, the legless, embryonic-looking tadpole evolved from the more "grown-up" Paleozoic hatchling, but fossils don't tell how. Fossils don't tell how frogs or salamanders evolved from more primitive amphibians either, which may be because their small bones don't fossilize well, or because less energy has been spent in finding frog fossils than other kinds. We find early amphibians of surpassing evolutionary importance not because they gave rise to frogs, but because they gave rise to us.

Whatever its origin, the tadpole is clearly a very efficient way of dealing with evolution's appetite for death. Why waste legs on hatchlings the vast majority of which will sink swiftly back into the ooze from which they sprang? Instead, let them be as humble in their needs as the ancestral worm, wriggling in the mud and scraping algae from the rocks, before the few lucky survivors turn into frogs and repeat the leap that the millennially fortunate lobe-finned fish made 395 million years ago. Evolution has no sense of history. It does not abandon past accomplishments to the fossil museum, but continues to play with them as though they'd happened yesterday.

REVENANTS

Insects are strong exponents of the proposition that if life can crawl out of primal ooze, it can just as well crawl back in. They have had a supremely successful time doing just that. Insects are even more apt embodiments than frogs of evolution's disregard for vertical hierarchies. Their young are more wormlike than tadpoles, but their adults have wings.

The first insects, known from Devonian fossils perhaps 350 million years old, were almost certainly land animals. They resembled springtails, wingless insects still common in the Klamath Mountains and elsewhere. At Rock Lake the surface above the crustacean alphabet soup was crowded with purple springtails as tiny as water fleas. While the water fleas hopped about below the water's surface, the springtails hopped about *on* it, as though in frustration at their inability to break through and partake of the oozy abundance underneath. Springtails gather in thousands on water or snow to mate, but they feed and lay their eggs on land.

The first aquatic insects were mayflies, which appear as fossils some twenty million years after their terrestrial forbears. Mayflies spend most of their lives in underwater, nymphal form, creeping about feeding on submerged vegetation. After a few months or years, they rise to the surface and become winged adults, which swoop about spectrally for a day as they mate and lay eggs in the water, then die. Adult mayflies don't even have functional mouths.

There is never a warm-season day around Klamath lakes when some mayfly species isn't sending forth its adults to hover in the late sun's rays, dimple the water with trout strikes, and attract bats and swallows that flutter over darkening water just as small pterodactyls must have fluttered over mayfly-rich lakes of the Dinosaur Age, when mayflies were already 250 million years old. Time seems to run differently when mayflies are swarming, as though the earth were closer to the sun.

But it isn't only the primitive mayflies that have reverted to the abundance of ooze. The true flies don't appear as fossils until the Jurassic period, when mammals already existed. (Mosquitoes and horseflies never would have evolved if their victims hadn't come first, obviously.) True fly larvae are even more wormlike than mayfly nymphs — grubs and maggots and wrigglers with only vestigial legs. Yet these creeping larvae have somehow provided the true flies with a reproductive potential unmatched on the planet. Their success in overrunning the earth is gauged by the difficulty we have in imagining a world without houseflies, mosquitoes, gnats, and thousands of other buzzing, biting insects.

Some of the worst tormenters and decimators of the human race have been aquatic true flies. The relationship between them and various deadly fevers, only very recently discovered, has justified ancient fears of still water, although the real monsters to be feared were microbes, not evil spirits. Aquatic flies are perhaps the most formidable organisms that the formidable arthropod phylum has produced, not only in their bloodsucking, prolific way of life, preying on animals thousands of times their size, but in their conquest of every freshwater niche. They thrive equally well in the most polluted and pristine of environments. Anyone who has been in a wilderness area has felt their power; only physical elements such as cold, heat, and rain have a greater effect on vertebrate life.

Klamath Mountain flies don't get thick enough to make breathing difficult as do those of tundra and muskeg, and they've never been a major disease vector like the Sacramento Valley *Anopheles* mosquitoes were in the 1820s when, infected with malaria by fur trappers, they decimated the Valley Indians. The trappers called the Sacramento the "valley of death," unaware that they themselves were the exterminating angels. Klamath flies still can be pretty disagreeable. I've been needled by no-see-ums along Clear Creek; clumsily stung by black flies under Preston Peak; buzzed and bled by purple-eyed horseflies at Babyfoot Lake; and monotonously teased by small, reticent mosquitoes at Taggart's Bar.

Still, aquatic flies are keystones of elegant ecological structures: bird and fish feeders, wildflower pollinators. One of my favorite wilderness animals is the chironomid midge, or gnat, which serves as a water-quality engineer in its larval stage, then becomes a kind of entomological firework when it leaves the water as an adult — a combination of utility and beauty few organisms rival. Chironomid larvae live in tubes in lake-bottom ooze: they spin webs in the tubes and pump water through by undulating their bodies. Millions of larvae reduce the stagnation of deep, cold water, oxygenate the mud, and circulate nutrients for use by other organisms. After months of this worthy toil, the larvae pupate and go to heaven, so to speak, emerging from floating pupal cases as tiny, diaphanous flies that swarm above lakes every evening in the millions — throngs that catch the light, billow in the breeze, and otherwise behave as prettily as falling snow.

I remember the glittering midges, slanting sunbeams, and lace of pine, fir, and cedar tops at Meteor Lake. Backlit against the deep blue-green of the trees, the swarms incandesced as they rose into shapes like bursting rockets or huge, luminescent jellyfish, then vanished suddenly as they climbed above the tree shadows where blue sky showed through their translucent bodies. It was as spectacular as a will-o'-the-wisp, as though I were indeed watching exhaltations of ancient ooze creatures rising from the lake. In a way I was, since chironomid larvae live very much as the tubificid worms that must have inhabited Paleozoic ooze.

Life has never really emerged from ooze, it has merely put an arm and leg out. The most celestial spires of civilization are as firmly rooted in it as the midge swarms. If the living ooze of lakes and seas were sterilized, it would be a black hole at the center of life, sucking everything back into death, because the elements of the biosphere that ooze originally produced — oxygen, nitrogen, proteins, hydrocarbons — are still essentially produced by it. The floating plankton that creates most of the world's oxygen is dependent on upwelling nutrients from the sea bottom. It is no wonder we feel a stab of anxiety when we find a mountain lake sterilized by acid rain from air pollution. It might as well be our own ovaries and testes that are sterilized.

Ecologically important as insects are, they are not the most important organisms to have reverted to the ooze. This distinction goes to plants, which evolved from algae through oozy species such as quillworts and ferns to the heights of flowering trees, then turned around and evolved back into plants that look like algae. I remember finding a greenish, floating weed that I would have called an alga if it hadn't raised two tiny yellow blossoms above the water. It was bladderwort, a flowering plant that has taken advantage of lake crustacean abundance by evolving tiny bladderlike structures on its stems that serve as traps, opening and sucking in unwary water fleas and copepods that trigger them. The crustaceans are digested to supplement the rootless, leafless plant's nitrogen supply.

Organisms as sophisticated as flies and bladderworts make a lake's air of primordial mystery something of an illusion. The methane that bubbles from a Klamath Mountain tarn is really no more primordial than methane generated at a sewage plant; they are both contemporary. Still, the human presumption that we are infinitely above the life of ooze perhaps is justified by the fascination and delight which this sense of distance, of safety, permits us. One of the great values of wild, remote places such as the Klamaths is that they let us look at the world not only as a workaday continuum of reproduction and death in which we are caught up whether we like it or not, but as a place of dreams, a place to lie on a sunny log and watch 395 million years of evolution rise toward us in the form of an ascending newt.

CHAPTER FOUR

RUNNING WATER

I do not know much about gods; but I think the river
Is a strong, brown god —sullen, untamed and intractable
. . . .
Keeping his seasons and rages, destroyer, reminder
Of what men choose to forget.

T.S. Eliot
The Dry Salvages

THE OBVIOUS NEXT STEP for life after primal ooze is onto land. Moving directly from lakes to forests in the Klamath Mountains, however, would slight a realm that always has linked ooze to land — running water, which is not water in quite the same sense as sea or lake. Seas and lakes are conservative, placidly sitting in their basins, jealously guarding the wealth of their depths. Rivers and creeks are radical — cutting new channels, throwing out sporadic bonanzas such as salmon runs — but poor compared to the old, established seas. The form of running water is literally radical; the networks which streams cut into land resemble the root systems of plants. As plant roots link soil and sky, so rivers form branching ways in which the life of sea and land move up and down.

Mythologies always have recognized the restless, anarchic nature of running water. Each river had its individual godlet, usually not an easy one to get along with. Lakes might be eerie, malign places;

but rivers could be actively dangerous, as with the Lorelei, female sprites that lured Rhine sailors to their deaths. The human-headed serpents that lurked in Klamath Mountain lakes also infested the rivers, and there are stories of giant footprints in sandbars, of hairy, humanoid swimmers who grab the nets and lines of fishermen at night. Female giants have been reported feeding on water weeds in midstream.

The sinuous curves of rivers must have seemed fundamentally unreliable to the patriarchal, sometimes misogynistic cultures of recent millennia. Always changing, flooding and going dry, they could not be confidently charted like the seas. But rivers were vital sources of livelihood for most cultures — fisheries, irrigation sources, travel routes. They couldn't simply be avoided as could mountain lakes; their real or imagined perfidies had to be tolerated, propitiated. I've seen the tricks rivers can play. When I worked on a tour boat, the river sometimes would trap us in night fog and keep our charter parties out until dawn; or it would trip us with new sandbars, sending picnickers head over heels on the deck. That was the placid old Connecticut in New England; Klamath Mountain rivers play much more roughly. In 1964 they rose in awesome floods that scoured their beds so violently that their banks remain rocky and treeless twenty years afterward. They do this whenever a combination of heavy snow pack and warm rains sends unmanageable quantities of water down their steep gradients. There are highways in the region that were eight feet underwater during the floods, giving odd perspectives on civilization's penchant for building "permanent" roads and cities on one of the least permanent earthly realms.

The mythological uncertainty of rivers has persisted in evolution. Questions as to how, when, and why life moved out of the seas and up the rivers are less easily answered than those concerning land. There is no clear model of the sea-to-river transition corresponding to the water-to-land change of lobe-finned fish to amphibians. Moving up rivers may not have been as strenuous as taking to land, but it must have been hard to exchange the well-fed stability of the seas for the floods and turbulence of streams. Even more than the land transition, river evolution occurred as a case-by-case affair rather than a historical progression. Various creatures moved upstream, then down, then up and down again, while others invaded rivers from land or from inland waters. Some river organisms became so without moving at all, as seas or lakes in which they lived dried up, and the only remaining water lay in streams draining the new land.

Rivers are too volatile to be reliable sources of fossils. They don't simply fill in and leave large deposits of bones and shells and stems as do lakes and swamps. They often change course, leaving fossil river beds, but then another river comes along and erodes away all the evidence. There are plenty of river-bed fossils, but they tell a much less coherent story than do those of lake or sea beds. Millions of years of uninterrupted swamp and shallow sea life can be traced in the Paleozoic coal measures of the Midwest, but I can't think of a similar example of river life.

Almost of necessity, then, an approach to river evolution must be mythological; there is not enough evidence to support scientific models. And there is something very wonderful about rivers, slender ribbons of water stretched across thousands of dry square miles, bearing in their constricted channels life quite foreign to the surrounding land. It always has seemed miraculous to me to find some great fish suspended in a pool hundreds of river miles from the sea, as though it were one of those prehistoric creatures found suspended in amber. River life has something of the vividness and largeness we like to imagine in prehistoric landscapes, everything bigger and gaudier than the present life forms, as though the old, gigantic, profligate life of bygone eras will somehow cut its way back to us no matter how many miles of mountain and desert we put between ourselves and mother ocean. I've had dreams in which gray industrial rivers beneath rusted iron bridges and fire-gutted skylines suddenly have been brought to life by invasions of fish so big, diverse, and colorful that I was frustrated at the shiftiness of dream imagery, which wouldn't let me get a close look at the creatures.

Finding the Eel River summer steelheads in the Yolla Bollys was a dream come to life. It was September, the driest month, and it had taken two hot, fly-ridden hours to walk to the canyon bottom. I followed the bank upriver a little way, to a place where green serpentine boulders as big as houses were tossed around the river. There was a pool under a cliff — not a very deep pool, I thought, since I was able to jump right across the river at its outlet. Then my eye caught a flash of movement toward the pool's center. It was not the silvery darting of small trout or minnow, but a stately, gliding movement, unaccountably deep for a river I could cross in one jump.

I scrambled onto a boulder and looked into the pool. Suspended in it were perhaps fifty fish, each over two feet long. Some were so close I could have reached down and touched their backs. Others were near the golden gravel bottom, shadowy shapes, but magnified by the water's clarity to seem larger than those at the surface. I'd have thought the fish were salmon had not their speckled backs and red sides identified them as steelheads, the giant, ocean-going conspecifics of rainbow trout. There were a few rainbows in the pool, looking absurdly dinky and goggly-eyed beside their huge cousins. It seemed a steelhead might easily swallow one whole.

Someone had told me of bringing a politician's aide to see the summer steelheads as part of a tour of the river. The politico had become so fascinated with the steelhead pool that it had been hard to drag him away for the rest of the tour. I could see why. The pool had the preternatural quality of a work of art. It sparkled like a cut emerald in the sun-withered rough of the canyon, or not like an emerald but like some gem peculiar to the finned world, with facets in fishscale patterns instead of crystalline angles. The watery reflections on overhanging boulders formed fishscale patterns, the wind ripples on the pool's surface formed fishscale patterns — the steelheads themselves dissolved from sight into prismatic bars of fishscale patterns when the wind blew strongly. When the wind stopped blowing, the fish became visible in such detail that I could see individual scales on their backs sparkling like flecks of gold dust.

Flights of mayflies and dragonflies drifted over, red mayflies with golden eyes and orange or blue dragonflies. Occasionally, one would drop into the pool, and a steelhead would strike, sometimes merely thrusting its head out, sometimes leaping out bodily. It would fall back with a slap and splash that sounded more like a human's diving than a fish's. The steelheads had been living in the pool, eating nothing more than the occasional mayfly, since ascending the several hundred miles of the Eel from the Pacific the previous spring; and they would remain until the fall rains raised the river enough to let them swim to their spawning creeks — the creeks they'd been spawned in — farther into the mountains. After spawning, some might die of exhaustion, while others would turn and swim back to the ocean, growing even bigger on shrimp and sardines before ascending the river again to spawn in a year or two.

To find such oversize trout in such an undersize river made me wonder how evolution had brought about such a seemingly incongruous state of affairs. And that is not the most incongruous thing about Klamath Mountain rivers. If Klamath peaks stand on their heads, the major rivers, the Klamath and Rogue, seem to run backward — looking-glass rivers. While driving upstream beside the Klamath where it begins to rise from its deep, forested gorge into the bare buttes of the volcanic plateau, I began to see cormorants and sea gulls as though I were approaching the river's mouth instead of its upper reaches. If I had gone fishing in these high-plateau reaches, I might have caught bass, perch, or other fish typical of warm-water rivers. If I had gone fishing in the forested gorge, on the other hand, I would have been more likely to catch trout or salmon, cold-water fish.

This state of affairs — bass upstream, trout downstream — is the opposite of most rivers. Most rivers support cold-water trout in their upper, mountain reaches and warm-water bass in their lower, desert or valley, runs. The Rogue and Klamath seem backward in this way because they are probably older than the mountains they flow through to reach the sea — older than the Klamath Mountains. The deep gorges that both rivers have cut indicate that they already existed when the Klamath region was a rolling plain, that they gradually eroded the gorges as the mountains rose around them. If the Klamath and Rogue had been born at the same time as the present Klamath peaks, they would have had their headwaters in the peaks; instead, their headwaters are far to the east. They flow out of the white volcanoes of the Cascades, across the sagebrush desert of the volcanic plateau, then back into mountains, into the peaks of the Klamaths, before reaching the Pacific. The cold Cascades water warms as it becomes a sluggish, desert river, then gets cold again as it is fed by the Siskiyou, Marble Mountain, Trinity, and Kalmiopsis snows. The cormorants and gulls I saw so far upstream were not attracted by the ocean, but by the desert lakes of the plateau.

Other Klamath Mountain rivers, the Chetko, Illinois, Smith, Trinity, Scott, Salmon, and Eel (the Eel heads partly in the Yolla Bollys at the Klamath Mountains' southern tip though it mainly flows through the Coast Ranges), all originate in the Klamath Mountains proper, like normal rivers. But they share with the Rogue and Klamath another trait that may seem backward. Their clear, clean waters are poorer in life than desert rivers. More species of fish live in the upper, desert plateau region of the Klamath River than in all the rivers of the Klamath Mountains. About *five times* as many fish species live in a muddy little Ohio creek where I used to canoe than in all the Klamath Mountains. A hundred species have been recorded in that creek, Little Darby Creek near Columbus, while about twenty live in the Klamath Mountains, a region roughly the size of New Hampshire.

When one looks at Klamath Mountain rivers, though, their relative poverty of fish species is not so odd. Midwestern streams have been meandering for at least 300 million years over a landscape almost as level and stable as the ocean; they've had plenty of time to gather and hoard biotic riches. The steep gradients and boulder-strewn beds that make Klamath Mountain rivers so exciting for whitewater rafters also make life very hard for fish. Cold mountain water is poor in nutrients, and tremendous vacillations of water temperature and flow from rainy winters to dry summers cause great stress to organisms. Klamath rivers change from raging, icy torrents in winter to virtual strings of bathtubs in summer — neither is optimal for fish survival.

Despite their relative poverty of species, a great deal of evolutionary activity goes on in Klamath rivers, as in other challenging environments. I might call them evolutionary frontiers, although not in the progressive sense of the upwardly-mobile evolutionary myths. Rivers indifferently harbor some of the most primitive and most advanced of Klamath Mountain organisms, and some of the more primitive are better adapted to the rigors of river life than the more advanced, since they've had more time to adapt. All are evolving into the future, and the advanced organisms won't necessarily supercede the primitive ones.

SALT-WATER WEALTH

I may seem to exaggerate the poverty of Klamath Mountain rivers. As newspapers periodically report, they contain rich fisheries, rich enough for people to fight over. But they are fisheries of a special kind that amply partakes of the evolutionary ambiguities I associate with river life. They are anadromous fisheries, a term which derives from the Greek for "running upwards" and which signifies that the sought-after fish don't really live in the rivers, but merely run up them to spawn. The hundred-pound salmon or half-ton sturgeon that have been taken from the rivers (in the nineteenth century, when fish lived longer) did not reach their great size by feeding on river mayflies and minnows, but by gorging on Pacific herring, shrimp, crab, flounder, and shellfish.

The wealth of Klamath Mountain rivers is largely a borrowed wealth. They support their great anadromous species only when the fish are juveniles, and sometimes not even then. Although young steelheads feed in rivers, newly hatched Chinook salmon spend no more than three or four weeks in natal streams before they are swept seaward. When the mature salmon return to spawn in rivers after three or four years, they don't eat; and they die after spawning, the ultimate no-confidence vote for a river's nutritive potential.

Fish species that live permanently in Klamath rivers — trout, minnows, suckers, sculpins — are generally no more than a foot long. One of the larger species (and the only fish endemic to the Klamaths), the Klamath small-scale sucker, has a maximum length of sixteen inches according to scientific records. It takes a small-scale sucker nine years to reach a fourteen-inch length, not impressive compared to a salmon, which can grow that much in months. And suckers have the ecological advantage of being able to eat algae, which they scrape from rocks with their protuberant lips. If they were confined to animal foods — as salmon are — they probably would grow even more slowly.

The anadromous way of life is an ingenious way of balancing the poverty of river life against the abundant but competitive life of oceans. As eggs and fry, anadromous fish are cloistered in their sparsely populated spawning streams from teeming ocean predators, which even could include their own elders. Then, when large enough to survive in the ocean, they partake of a pelagic feast that lets them get much larger than they could in a mountain river. Reaching such great size and vigor, they are able to produce enormous quantities of young when they swim upriver to breed. One female sturgeon lays millions of eggs during mating.

These advantages perhaps explain why anadromous fish evolved; however, they don't explain how. Adult salmon didn't start swimming up rivers to spawn because they hoped to produce more young that way, nor did salmon fry start drifting downstream in search of more food. Nobody knows how the anadromous way of life started. Fossils tell roughly when various groups of anadromous fish appeared on the earth, but not much about how. We aren't even sure whether fish first evolved in salt water or fresh, although the general belief is that the first fish evolved in the seas, then moved up rivers, where they evolved into the modern, bony fish, which then reinvaded the seas and crowded out the primitive fish.

A difficulty with asking, let alone answering, these questions is that different kinds of anadromous fish have evolved quite independently. Klamath Mountain anadromous fish are from such disparate groups that they form a paradigm of fish evolution. The jawless, almost wormlike Pacific lampreys that swim upriver in spring to breed in small tributaries are directly descended from the ostracoderms, the first fish — in fact, the first true vertebrates. Ostracoderms were armored creatures that looked not unlike frying pans with eyes, snouts, and tails. Like lampreys, they lacked jaws and had tiny brains and cartilage skeletons. Unlike lampreys, they lived by scooping food from bottom mud instead of by sucking the blood of more highly evolved fish (no more highly evolved fish existed when they appeared, about 450 million years ago). Ostracoderms disappeared completely about 345 million years ago, and we have no way of knowing how lampreys evolved from them, but physical similarities make it seem likely they did. Immature lampreys still feed by filtering detritus from river mud before they return to the ocean as bloodsucking adults.

Lampreys might be seen as an evolutionary counterpart of the vampire myth. Like Count Dracula, they are the last survivors of an ancient race, anachronisms that feed on more modern beings. Adult lampreys lie in wait for passing salmon, mackerel, and other fish, attach themselves with their sucking mouths, rasp holes in the victims' sides with sharp tongues, and drink the blood and body fluids. Unlike vampires, though, they feed on a victim only once, and many fish survive the attacks. Lampreys have taken up this parasitical way of life only because more advanced fish have usurped the ecological niches their ostracoderm ancestors occupied: modern flounders and catfish now feed on the bottom ooze. And lampreys don't seem to do much ecological harm. Salmon and lampreys spawned together in abundance in Klamath rivers — like salmon, lampreys don't feed in spawning runs and die afterwards — and both

were major foods for Indian cultures, still are to some degree.
(They're commonly called eels in the Klamaths, although true eels
are bony fish not found in western rivers.)

Early bony fish that replaced the ostracoderms had much in common with the sturgeon that still run up the larger Klamath Mountain rivers. Sturgeon fin, tail, and jaw structures are like those of
300-million-year-old fossils. Oddly, sturgeon are even more primitive than the first bony fish in some ways; their spines are of cartilage, like a lamprey's, and their bodies are covered with bony plates
instead of scales, like the ostracoderms'. There's no known evolutionary explanation for such atavisms. Surprisingly little is known
about sturgeon despite their giant size and precious caviar. Nobody
knows how they spawn in Klamath Mountain rivers although they've
been seen jumping and otherwise acting sexually excited in late
spring. A jumping, twelve-foot sturgeon would have been an impressive sight.

The largest group of anadromous fish, the salmonids, doesn't
appear as fossils until the Cretaceous period, some 100 million years
ago. Nevertheless, salmonids are rather primitive compared to other
bony fish. Their fins are soft and positioned very much as are the
legs of salamanders, whereas more advanced types such as bass have
spiny fins set forward on the body, allowing better maneuverability.
Salmonid feeding and spawning also are unspecialized compared to
other groups. The salmon's anadromous way of life itself might be
seen as primitive, since relatively few of the more advanced groups
practice it. On the other hand, the scarcity of more advanced fish in
the anadromous niche simply may mean that salmon have occupied
it so successfully that others couldn't invade it.

Before people began blocking their spawning runs with dams, at
least, there was no doubt as to the salmons' success. Salmon packed
rivers like canned sardines; still do, in Alaska. I once stood above a
culvert near Juneau's city limits on the Fourth of July and was unable to see the bottom of a little creek for all the salmon in it. Such
sights are now rare in the Klamath Mountains; we are degrading the
salmon's habitat with dams, siltation from logging, and water diversions. A young man who lives in Hoopa along the Klamath River
says the river has visibly shrunk in his lifetime. I searched the Trinity
River for salmon one November and saw none. A motel keeper said
Indians had gill-netted them all; a gold miner said sport fishermen

had caught and illegally canned them all; a fisherman said he had seen two old ones. The river was opaque with rain, so there may have been salmon unseen, but not many. On the other hand, people who rafted the Rogue River in August of that same year found pools full of salmon, banks strewn with dead salmon. They got all the salmon steaks they wanted by cooking fish that died before their eyes. This wasn't normal; the salmon apparently had a gill disease and were dying before they could spawn. Even so, sick salmon are better than none, and the fact that the Rogue River is protected as a national wild river while the Trinity is dammed seems significant.

I've never seen a really big salmon run, but even the bits that persist in the Klamath Mountains have a deep resonance. I camped for a few days beside the Illinois River in the Kalmiopsis, one of the most untamed rivers remaining in the mountains, and every time a big fish passed, half visible in the swirling green water, it was as though it were dragging the Pacific behind it. The pools below the rock I sat on — wherein green algae waved like kelp, buttonlike brown algae covered every stone, and big orange crayfish crept over the bottom like lobsters — might have been tide pools.

Chinook, or king salmon, and steelheads are the main salmonids in the Klamath Mountains. The big Chinooks run far upstream, mainly in fall, while the smaller steelheads run in spring and fall. This sounds straightforward enough, but salmonid actualities are a bit more complicated. Even biologists can have trouble distinguishing fresh-from-the-ocean steelheads from salmon. It becomes easy once the steelheads have been in fresh water awhile and have acquired rainbow-trout coloration, but then it can be hard to tell steelheads from nonanadromous rainbows, except that steelheads are generally much bigger. Biologists have so far been unable to isolate any characteristic, except for their ocean-going habit, which intrinsically distinguishes steelheads from rainbow trout.

Salmonid evolution is knotty. They may have descended from small fish that lived in lakes and streams above the Arctic Circle when temperate forest stretched from Alaska to Greenland. Glaciation caused these freshwater fish to move into the ocean by burying more and more of their native streams under masses of ice, and by lowering the salinity of coastal waters with glacial meltwater. So the theory goes, anyway: there's not much fossil evidence. It is known that a fish much like today's Atlantic salmon lived a million years ago, and it is thought that the Pacific salmon species such as the Chinook evolved from it after having been cut off from the parent stock by Ice Age freezing of the Arctic Ocean.

Something about this puzzles me, though. The Atlantic salmon is classified in the same genus, *Salmo,* as rainbow trout and steelheads. If Atlantic salmon couldn't get to the Pacific, then how did steelheads? One of the more circuitous theories known to science attempts an explanation: a trout similar to the cutthroat trout (also genus *Salmo*) of the Rocky Mountains lives in Asia; and the theory maintains that an ancestor of this trout migrated to the Pacific coast (in the streams of a Bering Strait land bridge presumably), then ascended western rivers as far as the Rockies, where it evolved into the cutthroat trout; then the cutthroat trout turned around and migrated back westward, evolving into the rainbow trout, which evolved into the steelhead, thus completing a cycle for which the word "epic" seems pale.

There's an awesome dedication to ecological stability in salmonid evolution. Small fish living in fir- and pine-shaded streams of sixty million years ago took to salt water, made ocean-wide migrations, grew to giant size, ascended turbulent rivers — and for what? So they could continue to spawn in fir- and pine-shaded streams. I don't think any group of animals has shown a stronger attachment to the ancestral, north-temperate forest than the salmonids, even though the most they know of the forest is the taste of pine needles in the water and perhaps a glimpse of green boughs against the sky. Their evolution has been like a quest, in which desert, tundra, and ocean are crossed in search of the talisman of clear water running over clean gravel. We are affected when a dog travels hundreds of miles to go home; salmon have traveled millions of years. That they die after spawning makes the quest seem all the more heroic, and all the more tragic the possibility that the quest will be thwarted by dams which will silt up and become useless in a century or two.

DIVERS

For all their mysteries, salmon are one of the clearer strands of river evolution. We may not know how they reached the Klamaths, but we can be sure they came by water. There can be no such certainty with other strands, those that come from land or air or various combinations of air, land, and water. Insects are even more crucial to river ecosystems than to lakes, and they jump around dizzyingly among all three habitats. I was surprised at how many insects I turned up in the sand and cobbles of a river rapid: water beetle larvae, armored like trilobites; flattened, creeping mayfly nymphs; green and yellow shield bugs; fly maggots. White water understandably doesn't support the rich, floating algae and crustacean plankton of lakes, so the land-evolved crawling and flying skills of insects make them the foundation of river food webs.

Insects help to explain why turbulent Klamath Mountain rivers are able to support at least a few species of each vertebrate class, continuing the evolutionary paradigm begun with the anadromous fishes. This river paradigm may seem eccentric, since vertebrate evolution usually is viewed as a progression from water to land more than as a sequence of returns from land to water; but there's no denying the success of river amphibians, reptiles, birds, and mammals in their chosen habitat. From their viewpoint the goal of life is to get into rivers.

I found rough-skinned newts just as abundant in the Chetko River as in nearby Babyfoot Lake, although I thought the Chetko newts were twice as large as the Babyfoot newts until I picked one off the river bottom. It had looked a foot long underwater, such was the magnifying effect of the river's clarity, but it fit easily into my hand. Seen from the cliffs above, pools had appeared a few feet deep; when I waded in, I found them well over my head.

The Chetko in the Kalmiopsis is another place where a river approaches the wonder of dreams, not only in its waters but in its setting. As I waded along the bank, I passed under umbrella plants with leaves a yard across and found dozens of big orange and purple stream orchids in the thick sedges. The cliffs above were a rock garden of azaleas, rhododendrons, lilies, maidenhair ferns, and ankle-high patches of *Kalmiopsis leachiana,* the rare little plant for which the wilderness is named. With half-inch leaves and dime-size red flowers, *Kalmiopsis* is the oldest living relative of azalea and rhododendron, an almost extinct species that grows only on the Chetko, Illinois, and Umpqua drainages, and that wasn't discovered until 1930. The backpacker who'd penciled: "Turn back! Nothing here but mosquitoes and lousy fishing!" on the wilderness boundary marker hadn't known what he was missing.

In the springs and rivulets that make the Chetko cliffs so lush lives one of the largest salamanders in the world, the Pacific giant salamander, a stout, mottled creature that grows to a length of more than a foot and is known for almost doglike barking and snapping sounds. Like the newt, it is considered a primitive species because it still has lungs and breeds in water. Adult giant salamanders don't spend summers underwater as do newts, though, so they're less often seen. They lurk under damp stones or logs and are thought to breed in underground springs. Breeding may occur dozens of feet down, since one batch of giant salamander eggs was washed up as a well was being drilled. The larval salamanders, which have bushy gills behind their heads, are thought to wash out of the springs after hatching.

When returning from Preston Peak in the Siskiyous, still feeling dizzy after my bout of illness, I kept seeing little faces peering from pools in rocky rivulets I tottered across. The faces disappeared when I stopped to look, though, and I began to feel like Mr. Mole as he stumbled through the wild wood in *The Wind in the Willows*. Finally, one face kept still long enough for me to see the bushy gills behind it. But after I glanced away a moment, I couldn't find it again because its grayish blue and ochre skin blended so well with the rocks.

Pacific giant salamanders have been known to become neotenic, to breed while still in their larval form, never growing up to be terrestrial adults. Neoteny is an evolutionary phenomenon even harder to understand than symbiosis and preadaptation. There is a certain logic to the evolving of new life forms from amalgamation of two or more cooperating species, or from lucky possession of a trait that gives an unprecedented advantage when environment changes. But for evolution to occur by an organism's immature form taking over the reproductive role of the adult seems to go against common sense. That is what happens in neoteny, though. In fully neotenic species, the adult form disappears completely, and the species may take up a way of life quite different from that of its "mature" ancestors. Humanity is a neotenic species. With our sparse hair, small jaws, bulbous heads, and lifelong tendency to frivolity, we are more like the young of our primate ancestors than the adults.

Neoteny may be caused by environmental stress, to which organisms often respond by reproducing at an increasingly early age.

When mortality is high from such stress, individuals that reproduce early in life are more likely to leave offspring than later-breeding individuals. Their genes predominate, and the entire species begins to reproduce earlier. When secondary characteristics that formerly defined breeding status (for example, a full coat of hair, large jaw, and serious disposition) are no longer tied to reproduction, they begin to be selected out of the population.

Salamanders provide many textbook cases of neoteny. One example is the axolotl, a pale, neotenic species of Mexican lakes. Axolotls will turn into tiger salamanders — a black and yellow species common in the United States — if a certain trace element is added to the water they live in, which demonstrates that the axolotl evolved neotenically from the tiger salamander. Theory has is that a population of tiger salamanders in Mexico was driven into underground waters by increasing aridity on the surface and that this put some stress on the species that made them begin breeding in larval form.

Neoteny seems particularly suited to the stressful life of running water. Several North American river salamanders are completely neotenic, and one group, the sirens, has even lost its hind legs. Like rivers, neotenic evolution has an unpredictable volatility. It is a kind of eddy in the evolutionary current. When creatures evolve by becoming more like their own young, they seem to resist the progressive adaptation we associate with Darwinian theory. Becoming like one's young hardly seems the way to prevail in the struggle for existence. It seems backward, a retreat. Neoteny is not really a reversion, though. Neotenic salamanders may live underwater completely, like fish, but their bushy, external gills are quite unlike a fish's. A neotenic organism is something new, and thus may have considerable evolutionary potential, even though it may not be as well adapted to its physical environment — may not be as tough, quick, or strong — as its "mature" ancestor.

The otter is another Klamath river organism with neotenic traits, although its neoteny is less marked than the giant salamander's, and seems more behavioral than physical. Otters belong to the weasel family, by popular reputation the most savage group of carnivores, and one of the oldest. Otters share the extraordinary strength that makes weasels able to kill rabbits twice their size, but otters aren't nearly as solitary and businesslike as their relatives. They appear to be evolving toward sociability, curiosity, and playfulness, traits more typical of weasel-family young than of adults.

I was sitting beside the Klamath River one evening when, out of the corner of my eye, I saw two bumps suddenly appear and disappear in the water. Jumping fish? Then otter heads began popping out of the river all around me, not just two, but dozens, it seemed. They were only a few feet offshore, craning their necks at me and snuffling with obvious curiosity. I didn't move, but they must have gotten my scent for they all dived at once, emerging a few minutes later on the far bank. I counted them as they played along the shore —sliding down steep places, chewing on fish, mock-copulating— and finally arrived at a rough estimate of seven. The group may have been a mated pair with offspring, but they were all full-grown, so it was impossible to tell. If there were adults, they didn't behave any more "maturely" than the others. After twenty minutes of tireless play, the group headed upstream, rolling through the water like a school of little dolphins.

It's a mystery why neoteny favors particular types of organisms such as salamanders and, apparently, otters. It's hard to imagine a neotenic frog. What would it be, a tadpole with legs? I know of no such thing (which doesn't mean it doesn't exist). The common frog of Klamath Mountain rivers, the yellow-legged frog, is a normal, straightforward creature, sunning on the bank and bouncing into the water if disturbed. It certainly has no trouble breeding in adult form, though it must suffer a fair amount of stress in winter floods.

There is a frog in the Klamaths that lives almost entirely underwater, and it is a very strange frog indeed, although it's not neotenic. It lives in small, icy streams that drop down from the high country; and it likes near-freezing water so much that it rarely emerges into air, creeping about under submerged rocks instead, feeding on insect larvae. It dies of heat prostration if kept in water above forty degrees Farenheit. This is the tailed frog, a prudish name, since it refers not to a true tail but to a penislike extension of the male frog's cloaca with which he internally fertilizes the female. Typical frog mating, with the male shedding his sperm over already-laid eggs, wouldn't work in the tailed frog's habitat. The sperm would be swept away before it could fertilize the eggs. To further assure against this, female tailed frogs (which don't, of course, have tails of *any* description, making the name even more inappropriate) attach their eggs to rocks with a sticky secretion. Tadpoles have enormous, sucking, disk mouths with which they cling to rocks.

These ingenious adaptations to stressful white-water life might make the tailed frog seem highly evolved and modern. On the contrary it is one of the most primitive frogs alive, so primitive that it doesn't have the equipment to croak. It doesn't have organs for sensing airborne vibrations (it probably wouldn't be able to hear above the rushing water if it did). The tailed frog's only living relatives are found in New Zealand and are classed at the bottom of frog evolution.

Given its silent, submerged nature, I never expected to find a tailed frog in the Klamath Mountains. While pottering around the headwaters of Virgin Creek, though, I lifted a flat stone on top of a waterfall and caught a small, gray frog which, cautious examination proved, was indeed a tailed frog. Pictures I'd seen had made them look dull and ungainly, but this was a little gem of a frog. Its skin seemed set with tiny silver beads, and its eyes, which had vertical pupils like a cat's, were the color of garnet. The tail was shaped like a miniscule leaf and was a delicate rose color. It sat quietly in my palm until I splashed some water on it, afraid the heat of my hand might harm it, then it leapt into the water and disappeared under another rock, from which I was unable to extract it though I could feel it moving around. Turning over more rocks, I found tailed frog tadpoles — shiny black creatures with opaque eyes near the tops of their flattened heads. When I picked up one, it attached to my hand like a leech, although the tadpoles feed by scraping algae from rocks, not by sucking blood.

The tailed frog is not really rare; it lives in mountain torrents throughout the Pacific Northwest. Rather, it is an example of a creature so well adapted to its environment that it tends to disappear into it. There's a beauty in this harmony, but also a danger from the evolutionary standpoint. If the tailed frog is better adapted to mountain torrents than the galumphing yellow-legged frog, it also is less capable of changing, which is why there are relatives of yellow-legged frogs worldwide (the ubiquitous bullfrog is one) and relatives of the tailed frog only in the evolutionary backwater of New Zealand.

The plight of the tailed frog and its relatives is symptomatic of a major paradox in evolution. Natural selection moves toward adaptation of organisms to their environment, since those best suited to live in the environment are most likely to survive long enough to breed. If, however, an organism adapts too well to its environment, if there is very little conflict between it and the environment, and thus little pressure for it to change, the organism becomes increasingly vulnerable to changes in the environment. The potential for disintegration inherent in every environment can rapidly destroy the too exquisitely adapted organism, as the tailed frog would be destroyed if the trees that shade its streams were burned or cut down, letting in the intense Klamath sunlight to raise water temperature above its tolerance level.

Organisms need not be as ancient and inconspicuous as the tailed frog to be exquisitely adapted to their environment. The water ouzel, or dipper, is the only North American member of the passeriformes — the most modern bird order — to have adopted a complete river life. It can fly underwater, and feeds, like the tailed frog, by picking insect larvae from submerged rocks. Although fossils are lacking, ouzels are thought to have evolved during the glaciations of the last two million years, when the icy streams they like were common. They seem quite comfortable in water below fifty degrees Fahrenheit and don't migrate in winter unless their streams are ice covered.

While sitting beside the dank Trinity River during my fruitless November search for salmon, I was surprised to hear a bird singing as though it were a May morning. An ouzel was chasing another over the river, singing as it flew. The birds landed on separate rocks, like boxers returning to their corners, and began squeaking pugnaciously and tipping their back ends up and down. Another musical chase ensued. I haven't seen a more convincing demonstration of the theory that birdsong is used for claiming territory, not for attracting mates (November being as far as possible from nesting season). The second chase must have decided the property line, since the birds then went in opposite directions along the bank. One of them passed a few feet from me, unusually close, evidently too heated by the argument to care.

Ouzels don't seem to need a reason for singing, as though their satisfaction with their habitat transcends biology. When I was marveling at the summer steelheads in the Eel, an ouzel flew into a boulder cave overhanging the steelheads' pool and sang much of the time I was there. It wouldn't have been nesting in September, though the cave was the right place for an ouzel nest. I found a nest in the Siskiyous in as spectacular a setting as John Muir (who first discovered the ouzel's nest in the Sierra Nevada) could have wished for his favorite bird. Halfway up Clear Creek is Wilderness Falls, where a tilted megalith of granite forms a natural flume for the creek. I've never seen a falls quite like it. I was standing on the house-size megalith (idly imagining the exciting but possibly terminal ride the water might give me as it roared past and crashed into a deep pool below) when I saw an ouzel land on a cliff beside the falls. With an insect in its beak, it paused a moment before a clump of moss that was brown, instead of green like the surrounding vegetation. I surmised that this was the moss nest chamber which ouzels build to nest in, although there was no way for me to be sure by looking closer. The cliff was so slippery with spray that the nest was completely inaccessible except from the air. It's hard to imagine how an ouzel could have adopted a more aquatic way of nesting without actually laying its eggs in water.

Whether the ouzel's recent but ingenious adaptations to running water mean that it will last as long as the tailed frog has is an open question. It seems unlikely that there will be a shortage of snow-fed mountain streams in the foreseeable future. Conceivably, though, the climate could get milder and tectonic movements could stop uplifting mountains. Then the ouzel would be gone. No bird alive today is as well adapted to aquatic life as was *Hesperornis*, a loon-like, flightless fossil bird that had a flexible jaw for swallowing fish and feet so modified for swimming that it probably couldn't walk. But *Hesperornis* has been extinct for tens of millions of years.

Evolution is loaded with exquisite adaptations, but they are mixed blessings. Less ingenious designs often last longer. The western pond turtles that like to sun on logs beside the warmer stretches of Klamath rivers aren't exquisitely adapted to river life. Their closest relative is the eastern wood turtle, which rarely gets into anything wetter than a rain puddle. Yet turtles are the oldest surviving reptile group, very little changed from ancestors long predating the dinosaurs.

Evolution's favoring of the not-too-precise adaptation means that odd jumps such as preadaptation and neoteny often fit organisms for survival better than more businesslike adaptations. The otter, for all its "youthful" frivolity, is one of the oldest living carnivore species. It has existed in its present form for some thirty million years and inhabits rivers over most of the planet.

RIVER GENESIS

I had another encounter with otters, on the Eel River. There were three of them, an adult and two half-grown ones. I was conducting a fruitless search of sandbars for immature lampreys (which live in burrows like worms do) and may have looked a little eccentric with my pail and shovel, like a superannuated child wandered too far from a beach resort. As I bumbled across the narrow summer river from them, the adult otter became a little concerned, swimming up close to examine me. He or she evidently decided I was harmless, and the three climbed on a boulder and began scratching their bellies against the warm rock. It seemed such a casual, confiding moment that I felt I ought to make some attempt to communicate with them, some friendly gesture. But then I decided this might not be such a good idea. Who was I, a man, to be exuding beneficence toward wild animals?

In a sense we are no more ready to communicate with otters than they are to communicate with us. Humanity, with its connotations of sociability, curiosity, and playfulness, is tenuous in its applications to *Homo,* as it is to *Lutra.* Symbiosis may be crucial to evolution, but it doesn't happen every morning. And who is to say that species don't benefit as much from their conflicts as from their harmonies, if perfect harmony destroys the organism's ability to survive change? The vitality and beauty many people value in creatures such as otters arise in great part from their wildness, their apartness from the human symbiosis of domestication.

So I behaved toward the otters as I might have behaved toward unfamiliar people. I said hello and went about my business of digging holes in sandbars. The otters seemed to view this more uneasily than they had my mere presence. After covertly watching me fling sand around and make frustrated gestures for a few minutes, the adult edged the young ones away downstream, in a manner reminiscent of a nanny removing her charges from the vicinity of an odd-acting man in a city park.

It is hard to imagine where the evolutionary path of otters will lead. The existence of the sea otter may indicate they're moving toward the oceans, as did the dolphins' ancestors, sixty million years ago. Perhaps otters will become the dolphins of sixty million years hence. Nothing is ever quite the same, though, and otters perhaps have an intelligence and manipulative ability that the dolphins' ancestor lacked. The sea otter is the only tool-using marine mammal, opening shellfish by hitting them against rocks it holds on its belly. Tools, curiosity, playfulness, sociability — otters might be the "humans" of sixty million years hence.

The past and future evolution of water mammals is no less important than the evolution of land mammals. I said earlier that the evolution of river organisms seemed an eccentric paradigm for vertebrate evolution, but I'm not sure this is anything but anthropocentrism. River evolution may be a truer paradigm for evolution as a whole than the apparently more orderly and schematic land evolution. Like a river, evolution runs in fast and slow stretches, in floods and trickles, sometimes doubles back on itself, is at once destructive and creative. Science's vague picture of river evolution may reflect our true state of knowledge better than the seemingly more complete land model. And the flow of both rivers and evolution is threatened by the monument-building, geological aspect of civilization, which is wiping out large sections of both by attempting to channel their energies to feed its acquisitive desires. Damming a river and exterminating a species have much in common: both erase evolutionary futures.

Evolution would continue without lampreys, salmon, tailed frogs, and otters, of course. But river-evolved organisms may not all have been so "marginal." Indeed, if one theory is right, rivers produced the most "central" group of organisms in existence. There would be no world as we know it without flowering plants — no grass, no livestock, no fruits and vegetables. According to this theory, if rivers hadn't existed, flowering plants never would have evolved.

This may sound nonsensical, especially when I've been describing how Klamath Mountain floods scoured all plant life from their banks. How could such an unpredictable, stressful environment as a river have fostered the evolution of something as important as the flowering plant? Actually, we may never know how flowering plants evolved. There has been no solution to what Darwin called this "abominable mystery" because no fossils have been discovered linking angiosperms, flowering plants, to their ancestors. It's *likely* that flowering plants evolved from gymnosperms, the group including conifers, cycads, seed ferns, and other early plants, gymnosperms having appeared much earlier.

That evolution may have been a form of neoteny. Flowering plants have much faster and more efficient reproductive cycles than gymnosperms. Some botanists speculate that this greater efficiency came about when gymnosperms of perhaps 100 million years ago, that were living under very stressful conditions, began to reproduce in juvenile form. Juvenile features of these ancient gymnosperms then evolved by preadaptation into the characteristic leaves, flowers, and fruits of angiosperms. Some primitive gymnosperms were more like flowering plants than are today's conifers. Cycads had flowerlike reproductive structures and ginkgoes had broad leaves and fruitlike seed coverings. It is not hard to see how such plants might have evolved into true flowering plants.

But what stressful conditions caused this neotenic origin of flowering plants? Mountain and desert plants usually are so specialized for survival in their stressful environments that they don't reproduce well in gentler places. If flowering plants had evolved in mountains or deserts, how could they have conquered the moist lowlands they dominate today? It seems more likely that flowering plants first evolved in moist lowlands, but in some ecological niche that placed them under greater stress than their neighbors. One such niche is a river bank, which is a swamp part of the year, when the river is high, and a sandy or rocky desert when the river is low.

The first flowering plants may have been shrubby semiaquatics eking out a humble existence between the rivers of the Dinosaur Age and the lush conifer, cycad, and seed-fern forests. Their flowers and broad leaves may have been stunted, retarded versions of the showy reproductive structures and palmlike leaves of seed ferns or cycads, adapted to grab as much sunlight and produce as many seeds as possible before the next flood tore the plant's branches off, or the next drought withered them. Eventually, this speeded-up reproductive cycle gave these tough little plants an advantage over their slower-reproducing relatives. Then their spread away from river banks to dominate the surrounding forest was explosive, as indicated by the sudden appearance of flowering plants in the fossil record.

There's no fossil evidence of such a riparian genesis, and the most primitive flowering plants today don't grow along river banks, but in forest understories. (This simply could mean, however, that the first, river-bank angiosperms were later exterminated by their descendants, the willows and alders.) Still, neoteny is such a common factor in plant evolution that it's easy to believe the river-bank theory. The soft herbs and grasses that comprise the vast majority of flowering plants today are products of neoteny.

It always has seemed to me one of the great oddities of evolution that lowly weeds, grasses, and wildflowers should be the *descendants* of forest trees. It goes against the upward notions our culture subliminally instills in us. Things are supposed to keep getting bigger and higher. But it is fairly certain that the first flowering plants were woody instead of herbaceous. The gymnosperms from which they presumably evolved are all woody; furthermore, oaks and other flowering trees were among the *earliest* angiosperms to be fossilized. So it seems likely that the small, soft flowering plants so common today evolved neotenically as early, woody angiosperms in stressful environments such as dry or shady places reproduced in younger and younger form, until they became mere seedling versions of their barky ancestors.

River-bank flowering plant origins would give added substance to a river-based model of evolution, a model in which life bends and twists like a river instead of sloping upward like the land. The response of plants when faced with the obstacles of physically stressful environments is similar to that of rivers when faced with dams. They back up; they seem to diminish; but in that apparent diminution changes begin that eventually will circumvent the obstacle, perhaps utterly change the world.

Rivers and plants both bind the primal life of the seas to the continents. Plants are the strong scaffoldings on which the original symbiosis of photosynthesis has been raised from ooze and stretched as a canopy across rock bottom. Rivers are galleries leading the fundamental evolutionary potential of water throughout the land. If there is a point to all this architecture, it is not simply to move life from water to land, but to assure that water's creative flow and changeability continues in the stiffer medium of land.

CHAPTER FIVE

THE EVERGREEN FOREST

In the midway of this our mortal life
I found myself in a gloomy wood, astray
Gone from the path direct: and even to tell,
It were no easy task, how savage wild
That forest, how robust and rough its growth . . .

Dante Alighieri
Inferno

Now with serpents he wars, now with savage wolves, . . .
And giants that came gibbering from the jagged steeps.

Sir Gawain and the Green Knight

EVOLUTION ON LAND may not be the goal of life, but it's time I confronted it, though I confess to a certain reluctance. Stressful as rivers may be, there's an attractive sunniness to them, at least to the beach-loving modern mind. It's nicer to sit on warm sand, by sparkling water, than in moldy, mosquito-ridden forest shade. The common modern preference for shores is reflected in our enthusiasm for aquatic utopias, for living in undersea domes, talking with dolphins, milking whales. This would be a kind of fetal Eden, an ultimate neoteny in the original amniotic fluid of salt water. The evolutionary stages of lakes and rivers remind me of human youth. Lakes are like infants, lying in rocky cradles, full of tiny life. Creeks and rivers are like children and adolescents with their vacillations between slackwater apathy and flooding uproariousness. Klamath Mountain streams sound like crowds of school children — the same ebullient, nonsensical babblings.

If waterways mirror youth, then forests reflect maturity. Dry, complex, they stand above the waters as guardians and nurturers. When forests are cut down, silt chokes rivers and sunlight shrinks them. Trees are the pillars of evolutionary society. They don't babble like rivers: they whisper sedately in summer breezes or roar majestically in winter storms. Forests have their youthful moments, their spring flowerings, but they more often seem reticent, aloof.

There are no horizons in the forest. Beginnings and endings seem far away among the trees, and one feels caught in the fabric of life. This also is a trait of maturity, when the open reflections of youth fade before the confrontations of day-to-day existence. Getting lost in a forest is an archetypal fear: Dante astray in the wood, separated from the illuminations of birth and death, surrounded by the wolves and panthers of the flesh. One walks endlessly and arrives nowhere. I've lost a trail in the Klamaths a few times and can attest to the fear it invokes, even today, when roads are visible from almost all ridge-tops.

But we are land animals. The human mind is much more a product of forest shadows than of water's glitter. Perhaps this is what daunts us about forests, that they are like us, secretive, labyrinthine. We have lived with forest trees for millions of years, yet how little we understand of them. We only very recently have realized that, in their way, they are as alive as we are. If their lives seem rudimentary and simple, this only reflects the rude simplicity of our knowledge. It would be a greater thing than talking to dolphins really to understand the slow life of a forest tree as it passes its millennium of steadfast silence.

A big, virgin forest defies the intellect because of the deep interconnectedness of life in it. I sensed this in the twilit Siskiyou ridges on my first trip to the Klamath Mountains. None of the books or museum exhibits I'd seen had quite prepared me for those ridges, and I didn't have much sense, at that point, of "reality." It seemed possible that road work near Bluff Creek had been interrupted, as reports had it, by nocturnal visits from a being that left sixteen-inch, humanlike footprints in snow or dust, and that picked up pieces of heavy equipment and threw them in the creek. This being —called "Omah" in Klamath Mountain Indian language, "Sasquatch" in Pacific Northwest Indian language, and "Bigfoot" in English — is connected in Indian mythology with a variety of forest mysteries, from the calls of screech owls to spirits of the dead, "Omah" being a word applicable to many uncanny phenomena.

There are hundreds of reports of eight-foot-tall, hairy giants in Klamath Mountain forests, seen in flashlight or headlight beams, or in daylight. They have apelike or catlike faces, heavy jaws, and low foreheads. They have short hair — black, red, brown, gray, or even white — sometimes long and stringy on the head, like a goat's. They have massive muscles, including stout buttocks, and females have dangling breasts. Hunters tell of aiming at shaggy beasts, which then turn around and display these breasts. The giants walk like humans, but more gracefully, bending the knee as they put their weight on it. Their stride has been estimated at half a car length, and they bound across midnight roads with demonic swiftness. They make high-pitched whistling sounds. Besides giant footprints, there are reports of giant feces containing remains of water plants and eggs of parasites known from humans in China and the Pacific Northwest and of clots of hair unattributable to any known animal. An eight millimeter film allegedly taken near Bluff Creek shows a barrel-chested, hairy female striding across a timber-strewn clearing.

There is a consistency about most reports of giants. Some are spectacular: giants lifting boulders for a meal of ground squirrels or hoisting automobiles by their back ends. A few are preposterous: a man said a giant picked him up in his sleeping bag and carried him to a hidden valley, keeping him prisoner several days. The giant lived with a mate and two children, a menage that lacked only a family dog in its middle-class domesticity. (The captive escaped by feeding a can of snuff to the giant, making him too sick for pursuit.) But most giant stories are unassuming. Someone in the woods sees a large, shaggy creature picking berries or pulling roots and realizes with a shock that the creature is shaped like a human being. The creature looks up, sees that it is observed, and moves away. It does not flee in panic as a bear often will. It looks directly at the intruder, as a bear usually will not.

I've daydreamed many times of meeting a giant in the Klamath Mountains. It suddenly would be there, around a bend in the trail, or in the dimmest orange rays of my campfire. I might see the whites of its eyes as it glanced at me, then it would be gone. I've heard whistlings in the dark often enough, and not only of screech owls. I've been awakened, heart thumping, not only by whistles but by moans, gasps, shrieks. I'd get up and stumble into the trees with a flashlight. Often I'd see a mouse or a deer or two trees rubbing together in the wind. More often the sounds simply would stop, or ventriloquially fade away, down the canyon, across the ridge, into the shadow of overhanging trees.

What would we learn of forests if we had the senses of wild animals? The Indian idea of "Omah" in the forest, of things beyond normal human perception, expresses a real gap in our understanding. Wilderness is a great reminder of the limits of human perception. Where there are no clocks or roads, time and distance behave differently, like animals let out of the zoo. Where there are no signs or labels, things seem much less predictable.

People speak of their hair standing up when they meet giants, of an extraordinarily deep sense of fear, as though the ground had opened beneath their feet. I felt something of this as I stood before the pyramidal Siskiyou ridges on my first trip to the Klamaths. An entire, unknown dimension had opened suddenly, like a hole beneath my feet. I'd seldom felt so alone. Now, when I visualize those ridges, there is a dark shape against them, a human shape, but wider, in scale with the tree trunks.

It is a very old shape, by no means confined to western North America. The first English epic, *Beowulf,* concerns a giant named Grendel that might have stepped out of the Klamath night, except that Klamath giants are inoffensive in most reports, and Grendel is a man-eating killer. Grendel lives in a cave under a mountain lake full of serpents, and Beowulf, the monster-slaying hero, has to pursue him into the waters. An even older myth has so much in common with *Beowulf* — giant, hero, serpent, underwater adventure — that the two seem related. In Sumerian mythology, the first written mythology, a giant named Enkidu is the friend and protector of wilderness animals. Although Enkidu is a shy eater of wild plants, not a killer like Grendel, he also inspires great fear in humans, until he is befriended by the hero Gilgamesh. When Enkidu dies, Gilgamesh is so grief stricken that he dives to the ocean floor in search of the plant of immortality, which he finds, only to have a serpent steal it from him.

Most living societies have traditions of hair-covered but humanoid wild creatures: the "Yeti" of the Himalayas, the "Kaptar" of Central Asia, the "Almasti" of the Caucasus, the "Duende" of Latin America. I heard about the Duende in Honduras, where it was described as a little red-furred person who walked around with feet pointing backwards, making it impossible to track. Tales of captured wild men or women are common in Oriental and Western folklore. The captive is kept in a cage, trained as a servant, married to a human: it eventually languishes or escapes. Explorers' stories of such creatures met in out-of-the-way places are almost as common.

The prevalence of such stories suggests a physical basis for them. Fossils show that a number of humanlike creatures, hominids, preceded *Homo sapiens*. Hominids branched off from tree-dwelling, apelike creatures perhaps six million years ago; and they include our ancestors, although nobody is sure which hominids we evolved from. Early hominids did not have tools or fire, but lived by eating whatever small animals or plant foods they could find. They must have been very tough, resourceful animals, and it is at least possible that one or more species of hominid could have survived to the present, intelligent enough to avoid extermination or general discovery, retreating to more and more rugged terrain as human populations grew. A few fossil teeth found in Asia, teeth to fit a jaw larger than any living primate, suggest such creatures might be gigantic.

If such creatures exist, there is a certain logic to their existence. Civilization has explored the oceans, the icecaps, the moon, but it has not adequately explored the human consciousness. We may be able to predict the fate of the universe, but our own behavior continues to surprise us. So what wild animal would be hardest for us to discover? An animal very much like ourselves, perhaps. Certain Amazonian and Philippine tribes were not found until long after their regions had been "explored."

Such wild animals couldn't be *too* much like humans. If they were, they would betray their presence by competing for the same habitat. Could an animal be enough like us to escape our endless snooping, yet enough unlike us to escape our endless competitiveness? It seems a riddle for the sphinx. The forest's impression of evolutionary maturity comes to mind here. What if another hominid species had emotionally outgrown *Homo sapiens*, had not evolved the greed, cruelty, vanity, and other "childishness" that seem to arise with our neotenic nature? What if that animal had come to understand the world well enough that it did not need to construct a civilization, a cultural sieve through which to strain perception? Such a creature would understand forests in ways we cannot. It would be able to inhabit them even more comfortably and inconspicuously than bears.

There are places in the Klamaths where such a harmony between forest and hominid is imaginable, where the sense of aloneness and strangeness turns into something else. I found one such place a day's walk up the Chetko from strange, lonely Taggart's Bar. It was a dusty walk, through the dry chinquapin and tan-oak brush that covers large parts of the Kalmiopsis today. Much of the virgin forest was destroyed during mining days; ships' logs told of running along the coast for days and never being out of sight of carelessly or wantonly set fires. But occasionally one meets remnants of that forest.

I dropped into a little bench of Douglas firs and sugar pines so big that it took me a moment to adjust my perceptions to them. Some of the firs were more than eight feet in diameter, and it was disorienting to see them looming behind full-grown oaks and madrones that would have been large trees in another setting. An eight-foot-tall, hairy person would have seemed on the right scale for them.

I spent the night on a river bluff beside the grove and have rarely felt such tranquillity in a place. Sitting on a little promontory, watching nighthawks circle in a sky the color of star sapphires, I felt as though my veins and nerves were grown into the ground, connected with tree roots. The faint red glimmer of a madrone trunk in the dusk of the rustle of last year's yellow leaves falling seemed as much as I needed out of life. Not even the mosquitoes bothered me; I was woven into the fabric of root and branch. I had seldom felt so accompanied.

A VENERABLE UNITY

The virgin grove beside the Chetko was a good paradigm of forest evolution. As the firs towered almost three times as tall as the oaks, so gymnosperms have been on earth three times as long as flowering plants. Gymnosperm trees are petrified in 350-million-year-old rocks. Modern pines and firs are not as old as that; they seem to have evolved at roughly the same time as the first flowering plants. Pines and firs are adapted to mountainous, cold, dry, or fire-ridden places where angiosperm trees don't always thrive, so they have persisted while many more ancient gymnosperms have disappeared. Not much is known about how modern conifers evolved; they appear abruptly in the fossil record along with oaks, maples, and other flowering trees in the great temperate forests of the late Dinosaur Age, the forests of which the Klamath forest is a remnant.

Fossils do tell a great deal about these ancestral temperate forests, though, about their epic retreat from Dinosaur Age Alaska to today's temperate zone. We know how temperate forest displaced a subtropical forest of laurels, figs, cycads, and tree ferns in Idaho and eastern Oregon, how the subtropical forest retreated to the Pacific Coast (where it left behind the bald cypress fossils of the Weaverville flora), and how increasing dryness and cold finally drove the temperate conifers and hardwoods to the coast. We know that the ancestral temperate forest was more diverse when it first reached the coast than presently. There was more summer rain in the Klamath region ten million years ago, so trees now living only in the East existed there. Fossil forests include species such as chestnut, holly, elm, and red bay.

Summer rain dwindled with the rising of mountain ranges and the cooling of oceans during the Ice Age. Holly and chestnut disappeared, and many remaining plants switched reproductive cycles so that pollination occurred in earlier, moister times of year, or retreated to pockets where year-round moisture was available: stream margins, seeps, high altitudes with their heavy snow packs. Other species seem to have entered the Klamaths from the opposite direction as the climate grew hotter and drier. Live oak and madrone appear to have migrated north from central Mexico, where they still have close relatives.

These factors have left a complicated mosaic of forest in the Klamaths. A miniwilderness called Russian Peak just south of the Marble Mountains has seventeen conifer species in a square mile, a possible world's record: whitebark, sugar, western white, foxtail, lodgepole, Jeffrey, and ponderosa pine; white, Shasta red, and subalpine fir; Douglas fir (a different genus than true fir); Engelmann and weeping spruce; mountain hemlock; incense cedar; prostrate juniper; and western yew. Botanists are trying to make sense of this mosaic by sorting it out according to tree species that grow together. Unfortunately, they've come up with dozens of associations, and different botanists favor different associations.

To me there seem to be three basic kinds of forests in the Klamaths: a forest in which it snows little or not at all, a forest in which it snows a great deal, and a forest in which it hardly matters if it snows or not because bedrock is so inhospitable to tree growth that the vegetation is called forest with a degree of charity. The first forest grows below four thousand feet elevation, where winter precipitation falls as rain and where fog is the only summer precipitation. The second grows above four thousand feet, where winter snow reaches depths of twelve feet and there are summer thunderstorms. The third forest grows at any elevation where bedrock is serpentine or peridotite, rocks so poor in calcium and rich in heavy metals that no amount of rain or snow allows trees to grow in a more than sparse and stunted way.

All three forests are descended from the ancestral temperate forest. Fossils show that trees now confined to one or the other of the forests grew together in the easier conditions of the past, before the mountains had become so steep and the summers so dry. Of the three, the first type, the lowland evergreen forest, probably is most like the Alaskan forests of sixty million years ago. The grove I found on that secluded Chetko bench is an example of it. It is one of the greatest North American forests, its diversity unmatched in the West and exceeded in the East only by the southern Appalachian cove forests.

It is a five-storied forest. Douglas firs, grand firs, ponderosa and sugar pines, Port Orford cedars, and western hemlocks tower above; tan oaks, madrones, Garry, black, and goldencup oaks, golden chinquapins, bay laurels, and bigleaf maples form a broadleaf subcanopy; smaller trees such as Pacific dogwood, western yew, elderberry, cascara, buckeye, and hawthorn grow under that; a shrub layer of hazel, poison oak, wild rose, vine and Rocky Mountain maple, ceanothus, redbud, serviceberry, barberry, currant, gooseberry, blackberry, blueberry, huckleberry, salal, azalea, rhododendron, thimbleberry, salmonberry, and snowberry under that; and an herb layer of ferns, iris, violets, pyrolas, saxifrages, orchids, and hundreds of other wildflowers under that. In summer the wealth of deciduous plants makes it glow as green as eastern forests; and ferns, mosses, and conifers make it evergreen. It is stupefying to imagine a forest, of which this is only an echo, stretching from Alaska to Greenland. Exploring it would have made trekking the Amazon like crossing someone's backyard.

Tokens of this venerable unity are everywhere in the evergreen forest. Many Klamath shrubs and wildflowers are almost identical to those of eastern forests, despite millions of years of separation by deserts and prairies. Walking up Big French Creek north of the Trinity River in May, I kept passing plants that might have been blossoming north of the Ohio River, where I'd been the previous spring: false Solomon's seal, wild iris, wild ginger, yellow violet, redbud, dogwood. In the snowy meadows below Preston Peak in June, I found trilliums, fawn lilies, and marsh marigolds like those I'd seen growing near Columbus, Ohio, in April. Closer examination revealed differences that make the species separate, but considering the length of time they've grown apart, their basic similarity was impressive.

This similarity seems in defiance of natural selection, which we tend to think is producing a perennially different and improved world. But many forest herbs, including violets, false Solomon's seal, and trillium, have largely dispensed with the genetic recombinations of sexual reproduction, instead doing most of their reproducing by cloning new plants from creeping, underground stems. Such colonial sprouts are genetically identical to their elders, which explains why there has been little evolutionary change over millions of years. Relatively few new genetic combinations have been produced to compete with old ones. Most such plants still are capable of sexual reproduction; considering the attractive flowers they produce, it would be absurd if they weren't (although the showy flowers of violets actually are sterile, the plant producing tiny, green fertile flowers after the showy ones have wilted). But the tendency of many forest wildflowers to disappear permanently from places where colonies of them have been destroyed implies that the seed is no longer their major reproductive tool.

Cloning is not restricted to forest herbs. A great deal has been made of the age of giant redwoods, but those few thousand years may be only the tip of an iceberg of genetic invariability, since redwoods grow from root burl sprouts as well as from seeds. The ring of small redwoods one sees growing around a giant stump may be the latest incarnations of a genetic individual of incalculable age. Such genetic individuals may have been growing on the westernmost slopes of the Klamath Mountains since the species arrived on the West Coast. It's no wonder that fossil redwoods from the Dinosaur Age are so similar to those of the present.

Cloning is not exactly the opposite of neoteny, but it does seem to hint at a complementary tendency in evolution. Neoteny occurs where conditions are so difficult that an organism has trouble reaching mature form. Cloning occurs where conditions are so stable that the mature individual hardly needs to die, as on the floor of a forest that has changed little in sixty million years. Perhaps the distrust humans feel for the clone — the genetic individual endlessly reborn in exact similitude — shows how neotenic we are. We might like to live forever, but not in cranky, hairy maturity — rather in childlike innocence, in the prepubescence of angels. The idea of living forever with the ennui and neuroses of mature human personality is even scarier than that of dying.

It is suggestive here to note that no dead giants have ever been found. Perhaps the giants have approached the maturity of redwoods in some scarcely imaginable fashion, individuals being so hardy that they enjoy life spans beyond our comprehension. Such life spans presumably would generate the wisdom necessary to avoid being run over by civilization.

While this would be a bizarre adaptation for a mammal, it is a mistake to project our own genetic changeability as an evolutionary norm. For the "lesser plants" that form the indispensable matrix of forest life, change hardly exists. Many mosses, fungi, and lichens I passed while hiking along Big French Creek were not just similar to eastern species, they were the same. Some are distributed worldwide, which seems extraordinary for something as passive as a lichen or mushroom, not only in covering all that territory but in remaining unchanged in the process. What's more, some of these species may have persisted unchanged for millions of years. All this seems anti-Darwinian, since the plants have had so much space and time to evolve into new species. Wandering birds and mammals start evolving into new species almost immediately; indeed, their propensity for doing so is the foundation of evolutionary theory. If

Darwin had confined his observations to mushrooms instead of Galapagos finches, he probably would have ended as his family intended — a country parson.

Perhaps the uniformity of lesser plants over time and space is related to their main mode of reproduction, the spore. It makes them more mobile than they look. The cloud of brown dust expelled by a trodden mushroom or moss contains millions of spores light enough to float around the world on a spring wind. More important, the genetic material in a spore is the same as the parent's, making a spore-grown plant a virtual clone unless there has been some mutation. Even when fungi and mosses reproduce sexually, the exchange of sperm and eggs usually occurs on the same plant, limiting the potential for genetic recombination.

Given their limited sexual potential, I wonder how some of the lesser plants evolved the odd diversities they do exhibit, and what natural selection had to do with such diversities. I'm thinking particularly of mushrooms, the peculiarly varied spore-making structures of fungi. The actual "bodies" of fungi are tangles of threadlike protoplasm called mycelia that live in soil and other organic media. How did organisms so unvaried in form evolve such a bizarre array of colors, shapes, smells, and chemical properties in their reproductive parts?

I camped at a place below Marble Mountain where two robust fly amanita mushrooms had just raised their dark red, white-spotted caps above the soil. At sunset a lame doe went straight for the mushrooms and ate them. Had the mushrooms evolved as deer attractants? Would the spores mature in the doe's intestines, then find a rich medium for growth in her feces? Fly amanitas are very toxic to humans; but animals vary widely in their biochemistry, and the deer seemed unharmed the next day. What attracted her to the mushrooms? Deer are colorblind, like most mammals, so the showiness seems unnecessary, at least for deer. It would more likely attract insects such as flies, which are thought to spread fungus spores, although it is hard to see why an organism able to produce millions of spores that float on the breeze should need flies to spread them.

All this seems much less straightforward than a honeybee pollinating a flower. I don't see mushrooms competing for the attention of deer or fly as flowers compete for the honeybee's notice. Fungi seem peripheral to the struggle for existence, too amorphous in form and unpredictable in behavior for the straightforward competitions of the survival of the fittest. Mushrooms pop out of the ground in an abrupt way that appears magical, as though they are not subject to quite the same evolutionary constraints as seed plants are. There may be extinct and endangered mushrooms, but it would be hard to tell. A fungus can disappear suddenly from one place when it exhausts the nutrients in the soil, then appear just as suddenly in another place miles away, having grown from a single tiny spore carried on the wind.

That fungi don't fit easily into evolutionary theory doesn't make them insignificant. Their importance is central, particularly in forests. The earliest fossil land plants had fungal mycelia in their tissues, substantiating the likelihood that plants could not have conquered the land without the soil-building and cellulose-decomposing functions of fungi. The pines and oaks of modern forests depend on fungi that live symbiotically in their roots. The trees have lost their ability to absorb water and minerals directly with their roots; they get them from fungi. Perhaps these fungi began as parasites on tree roots, as many fungi still are, the trees gradually evolving poisons to repel the fungus invaders, the fungi then evolving resistance to the poisons, and so on, until mutual aggression subsided into accommodation, and the food drawn from the tree roots by the fungus more or less balanced nutrients and water drawn from the fungus by the roots. There's no way of knowing how or when such symbiosis evolved, but it could help explain the abrupt appearance of modern trees. The trees appeared so suddenly and have changed so little since their appearance, it is as though they are a new type of symbiotic organism.

Many common mushrooms are root symbionts; indeed, it is possible that most modern plants depend on root fungi to some degree. Some plants have even given up photosynthesis and become parasitic on fungi. The coralroot orchids I found so abundant in the Siskiyous get their food from corallike masses of tangled orchid and fungus cells that are their roots. With their spikes of waxy blooms, coralroots seem about as fungal as floral; perhaps they'll come to look like mushrooms in a few epochs. (Or, odd suspicion, is this what mushrooms really are — degenerate flowers?) Another common nonphotosynthetic plant, ground cone (which parasitizes madrone trees), resembles some purplish species of morel when it first bursts from the ground. It also looks like a fir cone, leading me to wonder if squirrels are thus tricked into spreading its pollen.

An uneasy sense that things are not what they seem always has pervaded the mythology of forests, and it is notable that it crops up in the evolutionary viewpoint. I see a likeness between the old, animist forest, where one could not be sure whether a screech owl's call came from a bird or an Omah, where there was no clear distinction between bird and Omah, and the evolutionary forest, with its unclear distinctions between tree and fungus, flower and fir cone. The tree-fungus relationship is as mysterious in its origins and implications as the owl-Omah one. Both belong to a world that goes deeper than appearances, where a buried interconnectedness of phenomena renders behavior ambiguous, where one cannot walk a straight line.

The individual's struggle for existence is a logical foundation for the theory of natural selection, but it doesn't entirely explain forest evolution. Individuals get so tangled with others that a cooperative evolutionary principle seems necessary alongside the competitive one. The evergreen forest has moved south from Pliocene Alaska as a unit, and it has been evolving as a unit along the way. It hasn't marched in orderly, parallel columns, each species traveling aloof from its neighbors, but as a mob, the ranks jostling and weaving, some marchers bearing crowds of others on their backs. As with a mob, the forest has been held together and propelled by forces we don't altogether understand, although we catch glimpses of them in things like cloning and symbiosis.

Individual species obviously continue to evolve by natural selection in the forest, as shown by the differences between fossil and living trees. But I cannot help feeling, on seeing a tan oak or chinquapin that is nurtured and watered by fungi, pollinated by insects, and propagated by jays and squirrels, that the evergreen forest is much more than an aggregation of competing entities. People often feel, on entering a forest, that they have encountered something with integrity and volition, with consciousness. I would be more comfortable about dismissing such feelings, which I have experienced, if we understood how our own consciousness arises from the tangle of neurons in our heads.

I said earlier that the human mind seems related to forest life. If so, I wonder what symbioses between them are buried in the ways we think and feel. Our awe in the presence of big trees suggests a deep connection. The evolutionary growth of the human brain was so rapid that there has been science fiction speculation that it was stimulated by some kind of fungal infection. Did we breathe our minds from the moldy air of a Pliocene forest? Symbiosis with fungi certainly is no stranger to humanity, not only in our pragmatic use of yeasts to make bread and wine. For many peoples the fly amanita mushroom I saw the deer eating was a divine food that connected with an invisible world where harmony between nature and humanity was maintained. Of course, bread and wine are sacred foods as well as staples. Fermentation may have been a greater discovery than fire.

CRAWLERS

The vitality of woody plants in the evergreen forest is overwhelming. Trees or shrubs take up virtually every square foot of space. Even after fires, charred oak and madrone stumps begin to sprout immediately. Their response to clear cutting the coniferous overstory is just as enthusiastic, prompting the timber industry and Forest Service to spray hardwoods and shrubs with herbicide so valued conifer seedlings will have an advantage. Herbicides could have unexpected effects on evolution in the Klamaths, since they affect genetic material. One effect would be resistance to herbicides, with the fastest reproducing organisms being the first to develop resistance. Since the "weed trees" foresters would like to eliminate are fast-reproducing, the timber industry might find itself wandering in circles eventually.

I doubt foresters will realize their dreams of an intensively managed, agricultural forest in the Klamaths — the kind of forest portrayed so attractively in magazine ads, with fawns and squirrels romping around neat rows of Douglas fir seedlings. Such forests work only on lands with agricultural potential such as the southeastern pinelands, which still can grow pines after their capacity to support tobacco or cotton has been exhausted. The evergreen forest has never been particularly accommodating to animal life, and there's no reason to expect it to be so for humans. Forest complexity has covered the continents since surprisingly soon after plants left the water; early club mosses and horsetails reached tree size. It is the form at which plant evolution seems to stabilize when conditions are optimal. If it weren't for fire, the whole planet might long ago have been bound in an impenetrable tangle of wood, in which animal life would have been much less important than it now seems.

In summer heat and winter damp one may walk the evergreen forest for hours and see nothing larger than a banana slug. Animals one does see often will be of the most primitive, crawling kinds. The first true land animals, the myriapods (centipedes and millipedes), are among the commoner creatures of today's evergreen forest. Myriapods look like armored worms and are linked to worms by a creature named *Peripatus* that lives in the tropics and is basically unchanged from 500-million-year-old fossils. *Peripatus* is soft bodied and spends its life in moist darkness like a worm, but it has a primitive version of the breathing mechanism that allows myriapods to live in dry conditions. (Worms aren't true land animals because they can't breathe unless their bodies are covered with a film of moisture.) Fossils indicate *Peripatus* lived in water as well as on land; perhaps it led an amphibious existence, like lobe-finned fish.

Somehow this wormlike creature evolved the chitinous armor and wiry legs of the millipede, which appeared in much its present form about 350 million years ago. Millipedes are thought to have evolved before centipedes because their fossils appear first and because they have two pairs of legs, instead of one, on each body segment. Simplifications such as a reduction of legs are considered progressive evolutionary traits, which has always surprised me a little. It seems uneconomic to evolve hundreds of legs in the millipede and then reduce them to dozens in the centipede. That millipedes are vegetarians and centipedes carnivores supports the land primacy of the millipede, however: there wouldn't have been much for a carnivorous myriapod to eat in the beginning.

Scorpions may have been the next to emerge from the water. Extinct scorpion relatives reached nine-foot lengths in Paleozoic seas. Klamath scorpions aren't dangerous, but are memorable anyway. When camped along the Klamath River, I was awakened by something on my face, brushed it off, admired the moonlight on the trees a moment, and went back to sleep. When I took up my sleeping bag next morning, I found a two-inch-long, purple scorpion under my pillow. It looked so flattened I thought my head had crushed it, but when touched, it raised its stinger and pincers and scuttled under my Ensolite pad. Evicted from there, it ran under my ground sheet, retreating under a log only after every bit of camping gear was removed — persistent creature.

The over-300-million-year persistence of myriapods and scorpions seems another exception to natural selection and the struggle for existence. Spiders and insects are so much more efficient that I'd expect them to have crowded out the earlier organisms, as the later vertebrates crowded out the lobe-finned fish and primitive amphibians. We look upon the invertebrate world as the epitome of coldbloodness, but from an evolutionary standpoint it seems more forgiving than the vertebrate. The many-segmented myriapod is not well suited to progressive specialization — it's hard to imagine a myriapod flying or building a web (they do spin little nests for their eggs, which they guard until hatching) — but this hasn't doomed the myriapods as primitiveness has doomed many groups of vertebrates. As long as there are fallen logs or rocks to crawl under, it seems, there will be room for millipedes and scorpions.

A group of vertebrates has become an evolutionary success by giving up one of its vertebrate traits and moving in with the millipedes. Unlike their amphibian ancestors, lungless salamanders breathe completely through their skin. Like myriapods, they breed and lay their eggs on land, the young passing through larval stages in the egg and hatching out as miniature adults. Turning over a log once, I found a lungless salamander curled around her eggs, while, a few inches away, a centipede curled around *her* eggs.

In the Klamath Mountains there are lungless salamanders with legs so tiny it takes a moment to realize they aren't worms: lungless salamanders that live forty feet up in trees; lungless salamanders with brilliant orange and yellow markings that come as a surprise in creatures that spend most of their lives under logs. It's no wonder medieval people thought salamanders magical beasts, born from fire. The belief perhaps arose from seeing salamanders emerge from hollow logs burning in fireplaces, but to me it accurately reflects the wonder of turning over a log and finding such colorful little beings. I found an individual with the lovely generic name of *Ensatina* every time I turned over its particular log for nearly a year, until it disappeared as though finally weary of my intrusions.

Lungless salamanders almost never come above ground except for a few nights during breeding season, and the ancient success of myriapods bodes well for their downward-mobile evolutionary choice. Of course, it's possible that myriapods and lungless salamanders are locked in a struggle for dominance of the under-a-log realm, that one group will eventually drive the other to extinction. I suspect, though, that there will always be enough slugs, springtails, and whatnot under logs to feed both centipedes and salamanders.

Almost all the dozens of lungless salamander species live in North American temperate forests. Their survival strategy is typical of the forest tendency of things to grow together. Lungless salamanders are much smaller than earlier amphibians, probably more timid and secretive too. They fit into the forest better than their ancestors, better than their living, lung-breathing relatives, the newts and giant salamanders. When forest ends in desert or grassland, lungless salamanders tend to disappear, whereas more primitive salamanders are more general in distribution. Lungless salamanders seem to have foregone an individual destiny in favor of the collective evolution I sense in mushrooms and orchids.

As with the tailed frog, there are dangers to this close adaptation to the environment. If temperate forests disappear, with their moist leaf litter and fallen logs, lungless salamanders will have made a bad bet. There may be a significant difference between tailed frogs and lungless salamanders, though. The tailed frog has largely adapted to a *physical* environment, the flow and chill of mountain streams, while the lungless salamander has adapted to a *biotic* environment, the complex diversity of the forest. The salamanders are tangled in roots, fungi, myriapods — one might say that they have thousands of allies in their struggle for existence. The tailed frog is also connected to many other organisms, of course; still, the tailed frog's struggle to feed and reproduce in its streams seems a more isolated one, and, thus, perhaps more precarious. One may get farther by hitching one's wagon to an ecosystem, however moldy and untidy, than to something as crystalline and pure as a mountain stream, or a star.

DRAGON GARDENS

The evergreen forest isn't always quiet and secretive. May mornings can seem downright bacchanalian with the flutings of western tanagers, black-headed grosbeaks, and solitary vireos; the drummings of blue and ruffed grouse; and the acrobatics and laughter of pileated and acorn woodpeckers. I vividly remember walking along Big French Creek and almost stumbling over a pair of copulating lizards, which I first thought were snakes. One hardly could find a more archetypal image of the earth's spring renewal; it was like finding two dragons entwined. They were each almost a foot long, big-headed reptiles called alligator lizards. The male held the female's head in his jaws, and they seemed in an erotic trance, hardly attempting to avoid me. The male was blue green, with a black and white checkered pattern on his back, showy colors I thought were a breeding display, since the female was brown, until I later found another pair of which the male was brown and female blue green.

Reptiles have been symbols of fertility since the earliest mythologies, probably because of their ability to shed their old skins and grow new ones as they grow larger. Mythical figures such as Oriental dragons show how they were venerated as ancestor spirits, grandparents of life. In prehistoric myths from Crete to Australia, the serpent was not the deceiver of Eve, of the mother goddess, but her consort, guardian of the garden rather than betrayer. Evolution, in a way, has vindicated these old myths by proving that reptiles really are our grandparents, that we could not have existed if they had not taken the next step away from primal ooze.

To walk up Big French Creek was to find ancient dragon gardens still intact. I came to a place called Cherry Flat, which might have been landscaped by forest giants, so preternatural were the arrangements of its wild plants: a rocky glade with beds of blue-eyed Mary, white nemophila, sulfur pea, and pink plectritis surrounding a tufted lawn of bunch grass and surrounded in their turn with tiny hedges of mountain mahogany, serviceberry, and redbud. On one side the garden overlooked a mossy, oak-grown cliff over the creek; on the other, it was overlooked by a gentle slope of Douglas firs. Cherry Flat was a masterpiece of wild flora, a place that seemed to transcend ecological categories such as forest or meadow and to have the uniqueness of personality.

The resident dragons of this garden were a multitude of small lizards that lived among the big rocks interspersing the flowers and grasses. They rivaled tanagers and grosbeaks in the gaudiness of their mating displays. When I sat on a boulder, a blackish fence lizard eyed me aggressively, pushing himself up and down on his toes and puffing throat and belly to show me the azure breeding color on his flanks and chest. Every scale on his back shone with coppery iridescence. Looking like a miniature *Tyrannosaurus rex*, he seemed ready to rear up on his hind legs and tear into me. Instead, he made a flying leap from his lookout rock and took off after a neighboring lizard, the two chasing dizzily back and forth before returning to their respective corners. The black tyrant puffed himself up again and did some more threatening pushups. Then he did what I would have done after such a ruckus. He closed his eyes and dropped wearily on his belly, suddenly deflating into a small, gray lizard, what fence lizards usually are.

Another lizard species, western skinks with azure tails, yellow-striped backs, and scarlet throats, flickered among the rocks, making it appear as though gas flames were being turned on and off. Less combative than the fence lizards, the skinks were even faster moving. Their scurryings reached a crescendo in late afternoon as the rocks baked in the sun, then stopped abruptly as evening shade fell. I assumed the passions aroused in pursuit were being consummated in lizard boudoirs.

The next day I met a more sober celebrant of reptilian fertility, as though the lizards had been merely comedic satyrs presaging the awesome revelations of tragedians. It was hot, and rattlesnakes had been on my mind as the trail skirted a rocky cliff. I entered a cool fir grove, saw a fork in the trail, stumbled from taking my eyes off where I was putting my feet, and heard the sharp "snick-snick-snick-sn-n-n-n-n" I'd been listening for on the cliff. The rattlesnake was on a log a foot below the trail, fully aroused, and ready to strike. I backed up, gave my heart a moment's rest, and took a second look at the snake. I could see the heat-sensing pits on the sides of its snout: like an extra pair of eyes, and somehow even more expressive of deadliness than its real eyes. They guide the strike. The western rattlesnake is not supposed to exceed four feet in length, but this one looked longer. At the time it looked to me like a tentacle reaching all the way from hell.

I wondered about the snake's excitement; I'd encountered rattlesnakes before without causing such hostility. Others had just lain there or crawled away without much indication they knew I was there, but this snake was not about to let me pass. I lobbed a bark chip at it and knocked it a few feet downhill. Its rattling grew even angrier and continued as I proceeded around a bend and met an even larger rattler stretched *across* the trail, not rattling, but in no hurry to get out of my way. I evidently had come upon a mating pair, or a pair of rivals, either of which could have explained the unusual aggressiveness of the first snake. I began to feel outnumbered, so I retreated (despite the expectant attitude of a hermit thrush that was scolding the snakes as though in hope I'd deal firmly with these potential nestling eaters).

I could understand why the old cultures revered the snake. Anything with that much power to dispense death could be perceived as having a complementary power to give life, an impression furthered by the snake's penislike shape and vaginalike swallowing abilities. In myths of early agricultural societies, the snake god who copulates with the earth goddess dies afterward (as ancient kings were killed after cohabiting a year with the queen) and is buried. His burial fertilizes the growth of food plants; in fact, they grow from his body, making him the source of agriculture, of the human symbiosis. He does not stay dead. He returns from the underworld in spring to begin the seeding and harvest cycle again.

The evolution of snakes bears an intriguing resemblance to the snake-god myth. Evidently, snakes really have gone underground and then returned to the surface, if not reborn, at least transformed. Snakes seem more ancient than lizards because they don't have legs — they seem closer to worms — but fossils indicate lizards evolved before snakes, and there is anatomical evidence that snakes evolved *from* lizards. Snake sexual organs and jawbone structure are like lizards', but their eyes are different. In conjunction with the lack of legs, this suggests that snakes originated as burrowing lizards whose eyes and legs degenerated from lack of use. There are burrowing lizards without eyes or legs today, and snakes with vestigial limbs. The males of rubber boas, primitive little snakes of the Klamaths, have clawlike appendages beside their sexual organs which they use to titillate the females.

My perception of the rattlesnake as a tentacle from the underworld was an oddly appropriate one. Snakes have won their considerable success in populating the modern world by first retreating to the underworld, losing sight and limbs as though maimed in some gory ritual, then emerging to evolve a whole array of new specializations — heat-sensing pits, detachable jaws, venomous fangs, new eyes. There's no doubt as to snakes' success in the Klamaths. On a fine spring day one meets a snake every few hundred yards on trails, usually small garter snakes, but sometimes a California mountain king snake, devourer of rattlers, a scarlet, buff, and black-banded beauty that would make a worthy bracelet for an earth goddess.

The small reptiles of the evergreen forest in spring are so colorful and lively that it boggles the mind to imagine how picturesque the ancestral forest must have been in prehistoric Alaska, when there were still dinosaurs. There may be more of this ancient world remaining in the Klamath Mountains than we once thought. Today's tanagers and grosbeaks may be descendants of small dinosaurs: new theories have challenged traditional notions of dinosaurs as dim-witted and slow moving and proposed that at least some dinosaurs may have been warm blooded, nimble, and fairly clever—like birds.

Some paleontologists are saying simply that birds are a surviving group of dinosaurs. The skeleton of *Archaeopteryx*, the first fossil considered ancestral to birds, seems that of a small dinosaur with unusual adaptations, feathers and wings. *Archaeopteryx* had teeth, like a dinosaur, and its wings probably weren't used for flying but for knocking insects out of the air as the creature ran around two legged in typical dinosaur fashion. (If so, the bird wing was yet another product of preadaptation.) Somehow, the feathered and winged dinosaurs survived the mysterious holocaust that exterminated their relatives. Perhaps their feathers shielded them from increasing cold, or from a deadly rain of cosmic radiation that sterilized their naked kin—so theories go.

Remove the feathers from a Klamath Mountain quail or grouse, and you have a fairly dinosaurian creature. Small dinosaurs may have been more endearing beasts, more quaillike, than we tend to think. Insect-and-fruit-eating species probably behaved with similar timidity and curiosity, perhaps clucking or cheeping to stay in contact with flock mates. Fossil evidence suggests that dinosaurs guarded their eggs and young. They almost certainly had gaudy breeding colors and may even have had territorial calls like birds. Grouse and quail have an aggressive streak reminiscent of conventional ideas of dinosaur savagery too. I was walking along the Illinois River trail in the Kalmiopsis when a feathered blur darted out of the underbrush making blood-curdling cries that startled me as much as the rattlesnake had at Big French Creek. It was a blue grouse, and it advanced on me more menacingly than any snake or bear ever had. I stood my ground, though, and it had second thoughts, finally strutting back into the huckleberries, its squawks changing in tone from bloodthirstiness to mere peevishness. If I had been three feet tall instead of six, I might not have come off so well in the encounter.

Predaceous dinosaurs may have had the graceful, noble air of today's hawks. They can't have been much more baleful than turkey vultures or than the owls that hoot and yap in twilit Klamath canyons. Certain dinosaurs may have been nocturnal hunters of primitive mammals; perhaps they hooted and yapped in prehistoric Alaskan canyons. A misconception arising from traditional notions of dinosaur gigantism and brainlessness is that mammals were inherently smarter creatures that fed largely on dinosaur eggs. Fossils indicate the first mammalian creatures evolved *before* dinosaurs, and mammals did not become dominant until long after dinosaurs (the nonfeathered kind, that is) had become extinct, which implies that the evolutionary superiority of *dinosaurs* kept mammals small, furtive, and nocturnal for about 100 million years. In the Klamath evergreen forest, where birds generally are more conspicuous and aggressive (at least in daytime) than mammals, the Dinosaur Age has never really ended.

The only mammals one is likely to see large numbers of in the evergreen forest are bats. After the lizards disappeared for the evening at Cherry Flat, dozens of tiny bats hardly more substantial than moths came out to chase insects in warm air rising from the rocks, often flying right over my head, steeplechase-style, so that I felt the breeze from their wings stir my hair. I fell asleep that night to the tiny rustles and clicks of bats catching mosquitoes a few inches from my face. Bats seem to live in an older world than most mammals; they do not have the wariness that makes their four-legged cousins so hard to see. (And, sometimes, so hard to live with. Bats never raid food supplies, chew holes in gear, or banish sleep with scoldings and clamberings in the underbrush.) In fact, bats are very old. Fossils show that they've changed little in forty million years.

Evergreen forest complexity imposes a strong conservatism on its animals. Small crawling or fluttering shapes seem best suited for threading the maze of the trees. Shrews, smallest and oldest of placental mammals, tend to be the commonest furry creatures of forests. Their way of life differs little from that of primitive mammals 200 million years ago.

The most primitive living rodent inhabits the evergreen forest. Moist places near water are honeycombed with burrows of the mountain beaver, or aplodontia, last surviving species of a family that thrived fifty million years ago, but later succumbed to competition from more efficient rodents such as rats. Aplodontias resemble short-tailed muskrats, but lack chisellike incisors and other modern specializations. The West Coast is their last refuge, and they're so shy that they're seldom seen. I sat all evening by freshly dug aplodontia burrows along Virgin Creek in the Trinity Alps but saw no movement. Such a tangle of ferns and wildflowers surrounded the burrows that the place itself seemed a primitive relic, an echo of gentler, lusher times lingering into the dry California summer. The flowers suggested quaint objects found in an attic: wintergreen and prince's pinelike wax ornaments, red-and-white-striped sugarsticks like old-fashioned candy canes, deep indigo monkshood, baby pink elephant's head, lacy white baneberry, vanilla leaf, and inside-out flower.

More modern creatures also fit into venerable molds. The almost-ghostly gray squirrels of the Klamath lowlands live much as did the first rodents, which were long-tailed tree dwellers. (Transporting and burying of seeds by squirrels and their ancestors probably was one of the ways by which the ancestral forest migrated from Alaska.) Black bears are among the most recently evolved carnivores, but they resemble fossil carnivores of forests fifty million years ago in their hulking, strong-jawed, sharp-clawed form, so well suited to living on roots, wild honey, and other forest foods.

The sameness of forest life can make it seem like a squirrel cage, one of those contraptions in which animals endlessly run, endlessly going nowhere. Maturity is like that, though, whether of individuals or biospheres. There are periods of sameness that can seem monotonous to a neotenic mind. Like Dante, we would rather travel through the seven circles of hell than continuously confront the forest shades of earthly maturity.

A great potential for transformation lies beneath the apparent stasis of forests, however, as it perhaps underlies the monotonies of human maturity. The changes are not highly visible; they tend to occur within the venerable molds of forest shapes, but they have made the forest a very different place from the way it was in prehistoric Alaska. We probably will never know what sounds greeted a May morning in that prehistoric forest, but we can be sure it wasn't calls of black-headed grosbeaks, solitary vireos, western tanagers, hermit thrushes, orange-crowned warblers, or others of the hundreds of songbirds that nest in the Klamaths. It's hard to imagine a spring morning without songbirds, but they are a fairly recent development, whether or not one considers them dinosaurs. They are at the forefront of evolution, evolving so rapidly that the family to which the orange-crowned warbler belongs is suspected of having produced a dozen new species in the past million years.

Recent as they are, songbirds have changed the forest dynamically, opening up thousands of ecological niches with their nesting and feeding habits. A tremendous increase in insect species from the Dinosaur Age to the present may be related to songbird pressure on insect populations. By geological time standards, the spread of songbirds has been explosive. Yet they seem the essence of the forest — the hermit thrush in particular, the brown, speckle-breasted bird that scolded the rattlesnakes at Big French Creek, seems the spirit of the evergreen forest even though the big trees have been evolving without the thrush's song for as long as they've evolved with it.

The evergreen forest is the best example in the Klamaths of evolution as accretion. Change in it does not occur as a revolution, but as an enhancement of familiar themes. The tangled complexity that makes it appear static gives the forest a great capacity to moderate change as well as produce it. One has only to climb above four thousand feet in the Klamaths to see how profoundly the world can change, and to be glad that the evergreen forest's venerable tangle has persisted in its accommodating primitiveness, its centipedes and salamanders, lizards and songbirds.

CHAPTER SIX

THE SNOW FOREST

So on he fares, and to the border comes
Of Eden, where delicious Paradise
Now nearer, crowns with her enclosure green
As with a rural mound, the champaign head
Of a steep wilderness, whose hairy sides
With thicket overgrown, grotesque and wild,
Access denied; and overhead up grew
Insuperable height of loftiest shade,
Cedar, and pine, and fir . . .

John Milton
Paradise Lost

THE FIRST TIME I climbed above four thousand feet in the Klamath
Mountains was a revelation. The lowland evergreen forest had not
seemed too unfamiliar to me. Its green, mossy density reminded me
of New England woods I'd known since childhood. The density of
the lowland forest along Canyon Creek was even a little oppressive,
and I dawdled at the edge of the wilderness for an afternoon, eating
raspberries and feeling unaccountable yearnings for parking lots
and cafes. I had never set out to walk across a wilderness.

When I finally crept a few miles up the steep trail, though, I
found myself in a forest completely new to me. A lifetime of scenic
photos had not prepared me for the real high-country forest of the

West. All broadleaf trees had vanished — oaks, madrones, maples — and I climbed through open stands of big ponderosa pines and incense cedars. The sunny slopes were covered with bunch grass and wildflowers; it seemed more like a park than a forest. It didn't have the shady, moldy redolence I associated with forests. It smelled aromatic, like vanilla or fresh pineapple. The air was so fresh that I could move from sweat to chill merely by stepping from sunlight to shade.

I didn't think about the sudden change then, I simply enjoyed it; but if I'd wanted to know why the broadleaf trees had disappeared, there was ample evidence in sight. Woody angiosperms hadn't vanished completely. I passed little clumps of broadleaved shrubs, healthy enough, but with a depressed, prostrate aspect, as though some large animal had slept on them. The "animal" was a very large one indeed. I had crossed the altitudinal boundary above which winter snow accumulates heavily enough to break branches of broadleaf evergreens more than a few feet tall. Freak snowstorms at lower altitudes demonstrate the power of snow over madrone and live oak branches; one can gather a winter's supply of firewood from limbs felled by one such storm.

The snow boundary varies in altitude according to slope exposure, but there is always a point above which tan oaks, live oaks, madrones, chinquapins, and bays don't grow. (Maples and deciduous oaks grow higher because they drop their leaves in winter, but tend to be scattered in moist, protected pockets.) At least, they don't grow as trees. Shrubby variants of some species grow far into the snow zone, where they huddle under the big conifers or form an impenetrable dwarf forest on dry slopes. They are another example of neoteny, forced to reproduce in sapling size by glacial conditions prevailing at higher altitudes. One species, Sadler's oak, which grows only in the Klamaths, has lost whatever large form it may have had, now growing only as a snow-zone shrub.

The amount of snow that builds up at high elevations in the Klamaths seems out of proportion to the mild lowland winter. Even in May the still-unmelted snowpacks of high plateaus loom like glaciers as one climbs the slopes toward them, and moving a few miles into such eight-foot deposits is like traveling back in time to midwinter. In such conditions I'm surprised not that the broadleaves are so small but that the conifers are so big. True, the red firs of the snow zone don't reach the eight-foot diameter of lowland Douglas firs, but they often reach half that size.

The red firs and mountain hemlocks of the snow forest are so well adapted to the glacial conditions above four thousand feet that they don't seem descended from the same ancestral temperate forest as the tan oaks and Douglas firs of the lowland, even though fossils prove they are. The snow forest forms a series of islands in the dark green sea of the lowland evergreen forest. This is graphically clear in winter when green canyons and valleys surround white peaks and plateaus, but it is equally clear all year in sharp differences in plants and animals above four thousand feet. The Ice Age hasn't really ended in the snow forest, and things have a kind of boreal vitality. The mammoths who left their bones near Douglas City inhabited an environment like today's snow forest.

Cold-loving trees that have disappeared elsewhere persist in the Klamaths. The forest around Babyfoot Lake in the Kalmiopsis is almost the only forest in the world dominated by the drooping branches and reddish, jigsaw bark of weeping spruce, which grew throughout the West once. Rocky ridgetops in the Salmon and Yolla Bolly Mountains support a scattering of foxtail pine, the only other groves of which grow in the high Sierra.

Snow-forest wildflowers display little of the pallid delicacy of lowland flowers such as toothwort or Solomon's seal. Everything is bright and tough: scarlet penstemon, shocking pink phlox, purple lupine, and incandescent yellow sulfur-flowered eriogonum. Often the forest floor is covered with bear grass, a lily-family plant with slender leaves resembling grass, but so wiry that livestock won't eat them. Every five years or so bear grass puts out a two-foot plume of white flowers that resembles nothing more than a floral roman candle, an apt accompaniment to the wildflower firecrackers going off around it.

In the evergreen lowland, leaves turn dull yellow or brown and sluggishly drop at the first heavy frost. In the snow forest, autumn peaks are covered with golden aspens and scarlet thickets of mountain ash, Klamath plum, chokecherry, and bitter cherry that attract noisy flocks of robins, waxwings, and band-tailed pigeons. Winter storms can be almost arctic, with temperature drops so sudden and severe as to bring to mind stories of mammoths found frozen in tundra soil, their mouths full of grasses and wildflowers as though some unimaginable glacial wind had frozen them as they grazed. Even in summer snow-forest trecs recede into a quiet muting of light that reminds me of June nights in the Yukon. I'm not sure what causes this duskiness, maybe just the shadows of ridges or the stiff screens of fir foliage, but it is palpable and unearthly.

Judging from the widespread association of gods with mountains, the unearthliness of high forest has been felt by all societies. Of course, it's convenient to assign gods addresses people seldom visit, but there is an Olympian quality about the snow forest. Everything seems pure and newly minted, as though the yearly blanket of snow crystals has purged all squalor and decay from the landscape. The unconfiding ruggedness of the terrain engenders a curiosity, a long-ing that draws people from ridgetop to ridgetop in half-conscious expectation that the next fir-shadowed hollow must hold something marvelous. And there always is enough of the marvelous to draw us farther, enough of the fragrance of melting snow and of an almost-musical silence to hint that if we can just walk far enough, past exhaustion and the fading of sunlight on the peaks, we will find the source of that longing, some place or being that will confirm the intimations on which myths are founded.

The range of the giants that reportedly left their footprints at Bluff Creek corresponds rather closely to the snow forests of the world, from the Klamaths north to Alaska and then south through the ranges of Eurasia. Giant, naked footprints in snow are the most definite signs the creatures leave. The first newspaper reports of abominable snowmen and yetis were based on tracks found in Himalayan snows by British explorers in the midnineteenth century. Even earlier, in 1811, a fur trader named David Thompson in the Canadian Rockies recorded a fourteen-inch track, which might have been a grizzly bear's since there were claw marks — "yet the shortness of the nails, the ball of the foot, and its great size was not that of a bear." The Indians with Thompson said it was not a bear's, but a mammoth's, track. What the Indians may have meant by "mammoth" 170 years ago is not clear.

If giant hominids live in the snow forest, our notions of primate biology will have to be revised. No primate known to science lives in a cold environment such as the snow forest without artificial pro-tection. The anthropoid apes are completely tropical, and there's no indication that humans were able to break out of the tropics until fire and clothing were devised, until we were no longer wild animals. Yet, if reports are true, giants live comfortably through high-altitude winters, even diving through the ice of lakes. Giants are not seen migrating to the lowlands in fall; some people have speculated that they hibernate, like bears, an adaptation that would be just as extraordinary for a primate as surviving subarctic winters without fire or clothing. Some Asian monkeys live in areas where it snows, but none are known to hibernate.

The Klamath giants have another trait extraordinary for a primate. They are fully nocturnal. Their eyes shine yellow, red, or green in a flashlight beam, indicating that they have a reflective tapetum behind their retina as do wolves, bears, deer, indeed, most mammals. But no known higher primate has eyes that glow in the dark. Gorillas build leaf nests in trees at night and are so reluctant to leave them in the darkness that they foul them with their excrement. Night blindness is an evolutionary disadvantage that was traded for the unusually precise (for mammals) binocular vision that allowed higher primates to first swing through the trees, then run across the open savannas. Everything about higher primates — reduced sense of smell, relatively dull hearing, lack of long whiskers or other tactile equipment — points toward a strictly diurnal creature. Yet forest giants bound across midnight roads as fast as horses, eyes glowing like lamps.

Few organisms deviate as widely from their close relatives as Klamath giants would appear to; science has not been able to get a grip on them. There are no specimens, no official documents, only an uneasy welter of speculation and hearsay that slops into the mass media from time to time. Civilized knowledge undergoes a kind of *petit mal* — an inconspicuous, indeed, unnoticed, lapse of consciousness — when faced with giants.

Giants may inhabit a darkness even wilder than the Klamath night. Obscure as the evolution of blood and flesh has been, that of the mind is virtually unknown. There is no ground to dig up, only dreams and feelings, although there may be fossils in these unstable strata too. Or, if not fossils, living shapes similar to ancient ones. As the dense evergreen forest produces small crawling and fluttering shapes, the mind may generate certain shapes by its very structure. The snake shape of fertility may be one such. It seems so archaic, so alien to the modern mind, a buried remnant of primitive reasoning, but it is still alive. An adolescent with no special knowledge of anthropology described this dream to me: He saw an enormous green snake that stretched to the horizon, as if coiled around the planet. Then, to his surprise, the snake broke apart into thousands of people, who sat down at an endless banquet table that had been the snake's body. So, that's how it eats, he thought, with wonder.

Hairy giants may be shapes like the snake. We pursue them as though they were. There are crude wooden statues of giants throughout northwest California, and I wonder how much difference there is between the impulse that carves them and the one that carved much older statues in anthropology museums. The conscious intention probably was different, but these shapes don't live in conscious intentions. I certainly was drawn to the Klamaths by stories of giants, not because I believed them or even because their veracity seemed important, but simply because they were alluring.

If the giant is such a shape, though, its significance is much more obscure than the snake's. Although it has existed in myths for as long as there is record, it has never had the centrality and clear metaphorical resonance of the snake. The giant has been a shadowy figure hovering at the edges of those human dramas in which the snake has taken such an important part — god, dragon, tempter, savior. The anonymous, Christian author of *Beowulf* identifies the giant Grendel as a descendant of Cain, a demonic cousin of humanity, but doesn't explain how Grendel (and his mother) happened to live in a Scandinavian lake and doesn't explain how Grendel fit into the pagan mythology that originated the epic. We may never know what giants meant to our ancestors. As with forests, myths may undergo great changes while retaining their ancient contours.

It is suggestive to me that the giant inhabits a place very attractive to the modern mind. Grendel's *mere* was thoroughly nasty by Anglo-Saxon standards, but the forest where we find tracks in snow is like paradise — pure, flowery, luminous. Of course, the snow forest is bitterly cold in winter and eerily silent at night, but that doesn't deter us from crowding into it in ever-increasing numbers, from reverently preserving it in national parks and wilderness areas while other ecosystems just as fragile (and more amenable to economic exploitation, it's true) are left to fend for themselves. It's as though we are seeking there some part of ourselves.

The Anglo-Saxons may have disliked mountain lakes, but they

felt more at home in the natural world than we do. For them the human body was a microcosm of the natural world; the four elements of air, water, earth, and fire were embodied in human blood, bile, phlegm, and lymph. Their world was full of useful correspondences between human and nonhuman. A plant shaped like a particular organ or part of the body could cure the diseases of that organ or part.

Science has pried apart these correspondences, and we no longer have the comfortable belief that the world was made for our use. But there is an abiding human desire for linkage with the life-giving forces of nature. If we cannot get everything we want from the forest any more, it would be reassuring to think that there lives a being that still can. It would be like having one's grandparents always alive, primal sources of wisdom and security. The Enlightment thought it had found such a being in the paradisiacal wilds of North America or perhaps Polynesia, but history disabused them of this, and anthropology has demonstrated that there is nothing very ancestral about living "primitive" societies, that they are as complicated as the "advanced," merely less civilized. What if some close evolutionary relative of *Homo sapiens* still lived, though, a being smart enough to evade our explorations, furthermore, a being strong enough to survive the most remote, savage, and ethereal of environments? What a strong link that would be to this frightening but powerful planet, to this interminable snake that feeds everyone.

YOUNG WORLDS

One of the reasons the snow forest attracts us is that it contains more lively, noisy mammals than other places. Reptiles and birds may be colorful, but we respond most immediately to the bushy-tailed cleverness of our closest relatives. I've walked through the lowland evergreen forest for days and not seen more than a gray squirrel, but I seldom walk through the snow forest for even a morning without meeting vociferous chipmunks, golden-mantled ground squirrels, and chickarees. These creatures don't disappear spectrally as gray squirrels often do, but stand up and scold me, in no doubt as to who owns the snow forest. Even at night the whisperings of flying squirrels break the stillness. A flying squirrel once glided or tight-rope-walked to a stuff sack I had hung high between two trees and ate half a lemon crunch bar, which it evidently liked, since it came back the next night for the other half.

On my first backpack into the Trinity Alps I was almost as impressed by the abundance of black bear signs as I was by the boreal vegetation. In places the trail was virtually paved with bear scats (they'd been eating raspberries, just as I had). I've experienced days in the Trinitys when almost every meadow, grove, and thicket seemed to contain various size bears, sitting contentedly, Pooh-like, in the grass or crackling absent-mindedly over fallen branches so that the landscape pulsed with an ursine vitality that was at once joyous and daunting to me, since the only communication I realistically could have expected with a bear would have been over possession of the contents of my food sack.

Keeping control of one's supplies can be a struggle in the snow forest, and not just because of bears. During deer season in the Marble Mountains, I met two well-armed hunters who complained that deer had stolen all their bread the night before. They'd started to make sandwiches in the morning, but could find nothing but empty plastic bags with tooth marks. I sympathized with them, since I'd been kept awake most of the night before by a demented young buck that snorted and stomped around my camp, chewing my sweaty socks for salt. Even in daylight he'd kept it up, in concert with a thieving chipmunk; when I left my pack to chase the deer away, the chipmunk would raid my nuts and raisins.

All this competing for my food supply has given me the impression that the Darwinian struggle for existence is somehow more manifest in the snow forest than in the evergreen forest. Organisms seem more ruggedly individualistic in the snow forest. Each chickaree and chipmunk knows exactly where its territorial boundaries are, and is energetic about proclaiming them. Pines and firs stand in proud isolation, unencumbered by lowland underbrush. Even the snow forest's nonphotosynthetic wildflowers have a kind of cheerful assertiveness. Snow plant, a two-foot scarlet spike that hits the eye from yards away, seems more individualistic than the dull purple ground cone of the lowland forest. Ground cone tends to grow in batches, but snow plant often stands completely isolated from other members of its species.

The snow forest is newer than the evergreen forest, a product of glacial conditions that have prevailed for only a million years or so.

Perhaps its inhabitants seem more independent and competitive because of this, as New World colonists were at least overtly more competitive than their European counterparts. New worlds provide more space for individuals to compete in since there are fewer individuals. Where there is plenty of space, colorful, noisy, aggressive beings are more likely to leave offspring than quiet, passive ones.

I've just been saying, though, that there are *more* mammals in the snow forest than in the lowland. More mammals, perhaps, but certainly not more organisms. The lowland has more fungi, bacteria, protozoans, and invertebrates. Mammals may thrive better in the snow forest because they have fewer parasites and diseases there. Snowy winters are easier for many mammals to survive than damp ones. It is easier for them to hibernate when a burrow is insulated by eight feet of snow than when it may be flooded by lowland rains. Mammals that don't hibernate must keep a high body temperature through the hungriest time of year, and then the lowland damp can be more stressful than the high-country frost, which at least immobilizes some parasites. Lowland animals such as the dusky-footed wood rat suffer their greatest mortality during winter rains. I can see why mammals seem subdued in the moist lowland; it may be harder for them to live in than it was for the giant dinosaurs.

Chipmunks, flying squirrels, and other snow-forest mammals do live in the lowland, though they're less common there. Other mammals shun the lowland even though they could reach the tangled canyon bottoms from their austere ridgetops in an afternoon; they seem separated from it by an invisible evolutionary barrier. These include some of the most vigorously self-reliant species alive. The marten, fisher, and wolverine are relatives of the otter, but while the otter channels its muscular energy partly into play and sociability, these boreal species excel in solitary hunting. It is remarkable to compare the behavior of coyotes and bobcats to that of the marten. Like their domestic counterparts, coyotes and bobcats spend a lot of time simply loafing and fooling around. They are efficient hunters when in the mood, but otherwise easy going. Perhaps half the times I've seen coyotes or bobcats, they've been actively stalking or pursuing prey, and even then there's a certain frivolity. Coyotes will chase deer they've no chance of catching, just for the joy of running; and bobcats will stalk nothing, for the play of creeping and pouncing. But I've never seen a marten that wasn't seriously exploring its environment for food the entire time I watched it. Martens don't waste a movement as they patrol rockpiles for pika or trees for chickarees, and they cover an extraordinary amount of ground.

Muscular energy has reached an almost supernatural level in the wolverine, as legendary an inhabitant of the Klamaths as giant hominids. (More giant tracks have probably been reported in the Klamaths than wolverine tracks.) Resembling a cross between a skunk and a grizzly, the wolverine is arguably the most formidable living predator. Bears and mountain lions defer to it. It lives as far north as there is land, runs down prey in deep snow, and has jaws capable of breaking most things. The wolverine is to the otter as an eight-foot giant might be to a human being: a shaggier, wilder branch of the same evolutionary limb.

The vigor of snow-forest life isn't confined to mammals. Any species able to survive the winters seems to thrive, sometimes in extraordinary abundance. Where there are fewer species, there is more opportunity for a species to corner a food supply. Insect species may reach astronomical proportions because the plants they eat are more likely to grow in unmixed stands in the snow forest. I met one such species in the Trinity Alps. I was halfway up a steep ridge at nightfall, so I crawled into a dry gully to sleep. I awoke at midnight to the sound of raindrops, and had waking nightmares of a cloudburst-caused flash flood sweeping me all the way back down to the Trinity River, but when I turned on my flashlight, I saw that the "raindrops" were bluish, yellow-striped caterpillars dropping from overhanging ceanothus bushes. They were larvae of the California tortoiseshell, an orange butterfly that has been frozen into glaciers by the millions during its population explosions. The caterpillars had eaten the leaves off most of the ceanothus bushes I passed the next day (and I passed a *lot* of ceanothus bushes).

It's easier to see the survival of the fittest in all this than in the intricate parasitisms and symbioses of fungi and orchids. When an organism breeds as profusely as the California tortoiseshell, it clearly is a winner, at least temporarily, in the struggle for existence. How does a parasite win, though? A parasite is not a competitor, but a freeloader, with a vested interest in the evolutionary success of its host. If it kills its host, it goes hungry. A parasite thus tends to evolve toward cooperation with the host, toward symbiosis. In doing so, however, it begins to lose its separate identity.

The older a world becomes, the more competitive evolution seems to be superceded by the cooperative evolution I dimly perceive in the evergreen forest. As increasing numbers of different organisms are evolved and start relating in increasingly complicated ways, competition seems to fade into the confusion. Perhaps it is merely masked by diversity, and remains the underlying principle, but then why have so many organisms such as mosses and centipedes survived so long unchanged? It seems to me that evolution plays between competitive and cooperative poles. Cells without nuclei competed until they grew together to become cells with nuclei. Cells with nuclei competed until they grew together to become many-celled organisms. Many-celled organisms competed until they grew together to become . . . ?

That civilized humans find the snow forest so attractive says something about our position between the cooperation-competition poles. Civilization may be largely a product of symbiosis, preadaptation, neoteny, and such noncompetitive accidents, but it certainly is addicted to competition now. Our populations outstrip the California tortoiseshell's. It may seem paradoxical in this context that a legend of hominid giants living harmoniously with nature in the snow forest should have come to fascinate us. But the giants' harmony with nature has a good deal of the competitive about it. After all, they are not weak, humble creatures living in quiet dependence on a gentle ecosystem. They are bigger and stronger than bears, able to live through a mountain winter unclothed, run through the night unlit. In effect, they are lords of the forest, which is what many civilized humans would like to be too.

We want harmony with nature, but on our own competitive terms. We want the easy harmony of the winner, who has no trouble loving the losers. Evolution has demonstrated, though, that we are winners in a largely accidental sense, that the qualities which give us our present competitive edge arose not from intrinsic superiority but from fortunate interaction with the ancient, complex tropical ecosystems that produced us. Whether or not giant hominids now inhabit the snow forest, fossils indicate most ancestral hominids were four feet tall, implying that they had to do a lot of fast footwork to stay clear of lions and leopards. They may have dominated the savanna in groups, but not in proud isolation, as giants would seem to inhabit the Klamath Mountains.

Maybe I'm vulgarizing the giants by seeing them in competitive terms. As I said before, there's always the possibility they've simply outgrown our neotenic vanities, that they would no more dream of lording it over their own ecosystem than we would dream of eating our parents. The giants inhabit another forest in the Klamaths that does very little to support our competitive notions of mammalian hegemony in evolution. The lowland evergreen forest may not be very hospitable to mammalian dominance, but at least it has fostered *some* mammal evolution. This other forest hasn't fostered *any*, although it probably is just as old as the snow forest. In fact, this forest has fostered hardly any animal evolution at all.

CHAPTER SEVEN

THE RED-ROCK FOREST

There is shadow under this red rock,
(Come in under the shadow of this red rock),
And I will show you something different from either
Your shadow at morning striding behind you
Or your shadow at evening rising to meet you.

T.S. Eliot
The Waste Land

I REMEMBER WALKING through one of the lusher high-elevation forests in the Klamaths — white pine, noble fir, weeping spruce, Port
Orford cedar — and seeing it suddenly stop. Without warning I
emerged into an open expanse of red rock, sparse yellow grass, and
small, scattered trees. I might have come to the edge of a clear-cut,
but there obviously had never been any trees large enough to log in
the sudden opening, which stretched as far as I could see into a deep
canyon. It was as though I'd stepped out of the shade of a building
into the hot glare of a summer afternoon.

It was a little like moving from the deep shade of lowland forest to
the sun-spangled snow forest. The sudden vista was exciting, and
the colors of rocks and wildflowers were vivid. But it wasn't the
same. A sense of invigoration was lacking, the faint icy tinge in the
air that refreshes those entering the snow forest on even the hottest
of summer days. The red-rock expanse before me was not celestial;

on the contrary, there was something infernal about it. I felt I'd get baked as red as the rocks if I spent much time in it. I checked my water bottle.

The trees were species one might find in the snow forest — incense cedar and what looked like ponderosa pine — but they didn't soar. Squat and prickly as imps, they seemed instead to nod to the dark rocks into which their roots had scrabbled. They cast no shade. The wildflowers and grasses, too, were genera found in the snow forest, but there was nothing parklike about them. Straggling over the hot rocks, they were all flower and hardly any leaf, as though hurrying to set the next generation of seeds before the heat seared them away. I felt no desire to sit and rest among these plants as I had in the snow forest.

Here, a step away from the paradise of the snow forest, evolution had produced something close to our idea of hell. I might have walked into the mountains of *The Waste Land*, where "there is no water . . . but only rock." Landscapes as well as souls can fall into angelic or demonic modes: the Klamath Mountains are full of places like Devil's Backbone, Bake-Oven Ridge, Hell Hole, and Brimstone Gulch. Many of these names originated with the early miners, who seem to have had clearer notions of the hellish than of the heavenly.

As I walked farther into this strange landscape, though, I kept coming to little springs, bright green with sedges and other marsh vegetation. A hermit thrush sang in the canyon, proclaiming in no uncertain terms that this was *not* the waste land. These things reminded me that, set in the same climate zone as the lush forest it adjoins, the red-rock forest gets just as much rain and snow. If it seems a place for the damned, it is not from any celestial mandate.

This apparent semidesert that often is as soggy as a sponge is scattered throughout the Klamaths, and it is pretty much the same whether it grows near sea level or on high ridgetops. About half of the Kalmiopsis wilderness consists of red-rock mountains sparsely peppered with dwarfed incense cedar and Jeffrey pine (a tree that resembles ponderosa pine but has different ecological requirements). A half mile or so of red rocks and stunted conifers in the Siskiyous or Red Buttes might give the impression that this is merely an aberration of the Klamath's other forests, an underprivileged vegetation growing on soil too dry and rocky for the big trees. But the Kalmiopsis soon dispels that notion; hundreds of square miles of red rocks and stunted cedars there insist that this is a landscape in its own right. It is a landscape that evolved for different reasons and along different paths than the other Klamath Mountain forests, even though it shares many plants with them.

The incense cedars and Jeffrey pines are small and strange because they grow on peridotite and serpentine, rocks which have risen from the earth's mantle some sixty miles down, where heat and pressure create an environment suitably infernal for the most depraved of Dante's Florentines. These rocks, not any lack of water, cause the seared and blasted aspect of the red-rock forest. Composed of silica, iron, magnesium, and heavy metals such as nickel and chromium, they are chemically constructed so as to hold calcium in a tight molecular bond. Calcium is vital to the growth of plant stems and roots, so its scarcity has an effect similar to shortage of water. Plants grow among the red rocks, but only plants resistant to calcium starvation. Vegetation is sparse and slow growing, as in a desert.

Certain conifers, grasses, and evergreen shrubs are the plants most resistant to calcium starvation; and peridotite areas all over the world sustain a scrawny vegetation of conifers, grasses, and evergreen shrubs. The actual species differ according to geography, but in a way the Jeffrey pines and incense cedars of the Klamath Mountain red-rock forest have as much in common with conifers on peridotite halfway around the world as they do with Jeffrey pines and incense cedars growing on nonperidotite bedrock in western North America. They have been naturally selected to thrive on soil low in calcium, and so have taken a step away from "normal" populations of their species. They may have been evolving apart for a long time. Geologists believe that peridotite began to be exposed in the Klamath Mountains about twenty million years ago, which is about when the ancestral temperate forest began arriving on the West Coast.

The red-rock forest may seem hellish to us, but it is a refuge to its flora. Jeffrey pine, for example, is not suited to the lush coastal environment in which Douglas fir thrives. It normally grows on dry eastern slopes of the Sierra Nevada. Its predominance in the red-rock forest suggests that the Klamaths were as dry as western Nevada at some time in the past, that Jeffrey pine once was a major tree throughout the region. When the weather turned humid again, returning evergreen forest and snow forest would have pushed Jeffrey pine out of most of the Klamaths, except in places such as peridotite areas, where Jeffrey-pine populations adapted to calcium starvation were spared the competition of Douglas fir and white fir.

The red-rock forest may be vegetated largely by refugees from various evolutionary upheavals. A classic example of this has occurred in the last century. The sparse grasses that grow on peridotite where it is too infertile even for shrubs are among the last large expanses of native bunch grasses on the West Coast. Bunch grasses were the predominant native grasses because they survived rainless summers better than annual or sod grasses; their perennial roots and tuftlike leaves conserved moisture better. But in the nineteenth century, aggressive Old World grasses such as wild oats and cheat replaced bunch grasses in most of the West. Adapted to a stable environment, bunch grasses couldn't compete with the weedy annuals when the land was plowed and grazed.

On peridotite, though, bunch grass still prevails. Fast-growing Old World weeds need plenty of calcium to produce the super-abundant seeds with which they've dominated more fertile areas. They trip over their own efficiency when faced with calcium-poor soils, while slower-growing bunch grasses eke out a healthy, if sparse, existence among lethal-looking rocks that might have been belched from planetary innards the day before. I looked in vain for Old World weeds in the Kalmiopsis, but found more bunch grass species than I could count.

These grasses still looked infernal, though, for all I knew of their natural history and evolution. Instead of the dark green I associate with grass, they had bluish leaves, purple stems, and bright yellow seed heads. I couldn't get over the impression that I'd entered a landscape outside normal human ideas of life on earth. The smoky grayish foliage of Jeffrey pine and rigid, jadelike leaves of manzanita abetted the impression. It was as though growing on rock from the earth's deepest recesses had set the plants outside our sanguine historical view.

Stories of forest giants strike one differently in the red-rock forest than in more "normal" surroundings. They are said to exude an overpowering sulfur stench, to peer into windows at night and shake buildings. It was in the Kalmiopsis that giants are supposed to have sacked a mining town. If forest giants can express a human desire for linkage with wild nature, they also can express a fear that what lies at its heart, in its deepest, most secret parts — the night, the wilderness, the earth's core — is alien.

Ideas of normality are defined by the limitations of our knowledge. Forest giants may not be hominids or anything else we know. They've been reported in the vicinity of cold, moving lights. I've seen unaccountable lights moving over midnight wilderness on two occasions; they weren't airplanes or comets. We don't live in a world that's completely nailed down. We say that convection currents in the earth's mantle cause the continents to drift, but we can't get sixty miles underground to prove it.

Few places in the continental United States would be as likely a repository for the unknown as the red-rock forest. It is emptier of humanity than southwestern deserts or southeastern swamps; you can't ride a jet boat or dune buggy across it. I can't think of a better place for giants to exist outside our knowledge.

Even without giants the red-rock forest would confound our notions of normality. An organism that is seen by everybody who spends a little time in the peridotite barrens is just as strange as a hibernating, nocturnal hominid would be, stranger, because such a hominid would be acting like a member of its evolutionary kingdom, the animals, while this organism doesn't. It is a plant that acts like an animal, rearing snakelike from red-rock swamps to devour insects. I call it cobra plant because it has a bright green hood from which a forked, tonguelike appendage protrudes. Attracted by the bright hood and by nectar, insects land on the "tongue," crawl into a hole beneath the hood, and are trapped inside by downward-pointing hairs that guide them to the bottom of the tubular plant, where they fall into a pool of liquid, drown, rot, and are absorbed.

I often heard the frenzied buzzing of trapped flies when passing patches of cobra plants, which cover wet hillsides with close-packed, yard-high tubes made even more distinctive by their flowers and fruits, maroon and yellow objects that dangle from long, spindly stems. There's no more arresting sight in the Klamaths than a patch of cobra plants — the tubes are a fresh lettuce-green that radiates vegetable vitality. I half-expected the plants to writhe and hiss like man-eating flowers in jungle movies.

Cobra plant's closest relatives are the pitcher plants that live in eastern swamps. Like them, it has circumvented the nitrogen deficiency of swamp soils by evolving its leaves into a tubular shape that allows it to trap insects and, thus, get nitrogen from their decay. Cobra plant would look at home raising its green hoods among the alligators and bald-cypress knees of the Okefenokee Swamp, where I found pitcher plants as common as dandelions on a lawn. During the Oligocene epoch, as the fossil Weaverville bald cypresses tell us, the Klamath region resembled the Okefenokee. I suspect that an ancestor of cobra plant lived in those swamps of thirty million years ago.

If cobra plant lived in Oligocene Klamath swamps, how did it survive to vegetate today's mountainsides? It is the only pitcher plant relative west of the Rockies, and it grows only in northern California and southern Oregon. Pitcher plants need plenty of sunlight as well as wet ground, and the vast, soggy barrens of the red-rock forest are among the few places west of the Rockies where sunshine and plenty of moisture coexist. Calcium deficiency is less of a problem for plants when peridotite soil is wet, allowing a greater diversity than conifers, grasses, and evergreen shrubs where springs and seeps moisten it.

The miniswamps of the red-rock forest may be the last vestiges of a landscape that otherwise vanished millions of years ago. There are places in the Kalmiopsis, for example, a little headwater socket called Bailey's Cabin, where impenetrable tangles of cobra plants, azaleas, swamp laurels, Port Orford cedars and bewilderingly diverse lilies surround dark peaty pools, where the *absence* of alligators, snapping turtles, anhingas, and other Okefenokee denizens is surprising. I found not even a newt in the Bailey's Cabin swamp, though, and the seemingly lifeless pools looked bereft and forlorn.

This absence of characteristic animals in an extravagantly characteristic floral landscape is the strangest thing about the red-rock forest. We are used to thinking of the plant world as a passive support system for the more important feats of animal evolution that culminated in us, and it is disquieting — even for a plant lover — to see an entire flora wandering off on its own without producing new animals. I wanted to pick up the perplexing stunted landscape and shake some unusual bird or lizard out of it, or at least a giant centipede. But the red-rock forest obstinately persists in zoological ordinariness (as far as we *know*). It is sparsely populated with the same chickadees and fence lizards that inhabit the other Klamath Mountain forests. Dozens of its plants exist nowhere else in the world, but none of its vertebrates are unique only to the red-rock forest. There may be more peculiarity at the invertebrate level, though I've seen little evidence of it. A tiny gnat lays its eggs only in the fluid at the bottom of cobra plants, its larvae feeding on the rotting insects there. There may be other special relationships between invertebrates and the thirty or so plants that live only on mantle rocks in the Klamaths.

Special relationships between insects and rare plants are not peculiar to the red-rock forest, though. The insects don't evolve in relation to the peridotite bedrock that gives the red-rock forest its special nature, and this raises a fact of evolution that is minimized or distorted in our "normal" view of it. Animals never evolve in direct relation to physical conditions such as peridotite bedrock. They cannot, because they are unable to make food from dissolved rock, water, and sunlight. While this fact is obvious to anyone who has taken high school biology, it is nevertheless deeply subversive to a view of evolution that puts animals in front.

My little *Golden Guide* to fossils devotes well over a hundred pages to evolution of animals and seven to plants. In a truly progressive view of evolution, however, plants would be given the greater emphasis, because they are the leaders. They always adapt to new climates, new soils, new terrain before animals do. Animals are unable to live in new environments until plants colonize them, unable to exploit new ecological niches until plants create them. Even the mechanical evolution called industrialization had to be preceded by the fossilized horsetails and algae that became coal and petroleum, and it is an open question whether this mechanical evolution will continue once fossil fuels are exhausted. It certainly seems unlikely that any great new advances in animal evolution will come about until plants evolve new ways of using sunlight, water, and minerals. Civilization would not have evolved without the preceding evolution of cereal grains from wild grasses. That agriculture was developed by cultural instead of natural selection doesn't make the plants less important. They may be sown and harvested by us, but they still do the real work of turning soil, water, and sunlight into food.

The red-rock forest seems much less abnormal in light of the plant kingdom's evolutionary primacy. Far from being a backwater of life, it is a landscape so progressive that it has left animals behind. It is dealing with a particularly knotty example of the basic evolutionary problem of turning rock into green growth. That animals haven't caught up with it is no reflection on the little incense cedars and Jeffrey pines and bunch grasses. They are doing their jobs, and if animals will be patient for perhaps another twenty million years, their descendants may benefit. Animals may be benefiting already in ways we don't know about.

Cobra plant begins to look even more impressive, a green serpent of a plant that devours flies in a landscape from infernal regions. It reminds me of the dream serpent that becomes a banquet, that feeds by devouring itself. Like other plants, cobra plant makes its own food; the flies are only a nutrient supplement, a botanical vitamin pill. Ancient myths merely reflect our instinctive animal chauvinism when they tell of food plants growing from a buried snake god. It is the snake that grows from the plant, via a mouse or grasshopper. Cobra plant is as close as I can imagine to a living embodiment of basic fertility, which must devour, must take life in, as well as grow, give life out. It is even perennial, the old hoods gradually turning deep red but remaining on the plant, so that a patch becomes a pipe organ of cobra shapes, from tiny new ones like fern fiddlenecks to brown mummies of hoods dead for years.

Like the snake, cobra plant and the other strange flora of the red-rock forest may seem, from the human viewpoint, to take a downward path in evolution, to move toward impoverishment and obscurity. If upward-minded people are attracted by plants, it is because they don't kill to eat as animals do; they seem less carnal, more ethereal. The gnomish conifers and creeping shrubs aren't very elevated, though, and predatory cobra plant can seem depraved and devilish. As land animals, from millipedes on, show, evolution creates by simplifying and reducing as much as by elaborating and elevating. The red-rock-forest inferno may prove more creative in the end than the snow-forest paradise. It is the obdurate physical adversity of things such as peridotite bedrock which often drives life to its most surprising transformations, as with the simplification and reduction of ancient conifers which brought about the flowering plant.

CHAPTER EIGHT

HIGH MEADOWS

I am the grass; I cover all
. . . .
What place is this?
Where are we now?

I am the grass
Let me work.

Carl Sandberg
Grass

ELEVATED AS THEY ARE, trees are not the culmination of plant evolution. This honor goes to the most simplified and reduced of flowering plants, the grass, which has evolved out of adversity. One hardly
would think, on emerging into one of the many small meadows
dotting the Klamaths, that one confronted the landscape that has
replaced forest over most of the earth, but it is so.

There are a few big meadows in the Klamaths. Morris Meadow in
the Trinity Alps was a culmination of snow-forest ethereality when I
reached it on my first trip. After four days of climbing, I passed
through an aspen grove and entered a corridor of tall grasses and
wildflowers so long I could hardly see an end. It was a relief to come
to such a soft, level place, so I camped there. When I took off my
pack, I felt I'd have to hang onto grass stems to keep from levitating

to treetop level — the turf felt like a trampoline. Twilight lasted longer than it had in the forest, as though the Sawtooth Ridge's rocks kept radiating stored sunlight after the last rays had left the sky. Robins still sang as the stars came out, and I could see the heads of deer grazing in the cow parsnips and corn lilies.

I felt I'd reached the top of the mountains, more so than if I'd climbed one of the nearby peaks. In an evolutionary sense I had. A mountaintop now is the same as a mountaintop four billion years ago — a piece of stone that rises or falls indifferently with the planet's tectonic stresses. But a mountain meadow, however small, is part of a botanical revolution that has changed the face of the earth in the past thirty million years. Fossils indicate that grasses were relatively uncommon in the Dinosaur Age and in the early Age of Mammals. Like the first angiosperms, they existed at the peripheries of the great ancestral forests and were forced to reproduce at increasingly small size by the adverse conditions to which forests consigned them.

But then the earth became increasingly dry and cool. I can imagine the fires that swept the ancient forests of Siberia and Alaska as periods of drought grew increasingly prolonged; single fires might have lasted decades, wandering across entire continents. As the ancestral forest proceeded on its long march to the western and eastern seaboards, the heart of North America was taken over by grasses and other small flowering herbs, whose neotenic ability to produce a new generation every year allowed them to thrive in a world swept by fire and freezing winds, a world in which the only shade was that cast by hurrying clouds or dust storms. It was an unprecedented world, and it had unprecedented effects on animals. The insignificant furry creatures that had haunted forests, little changed for a hundred million years, exploded in hundreds of grassland adaptations: the hoof, the ruminant stomach, the grinding molar, the giant herd.

To move eastward from the Klamath Mountains today is to relive this unprecedented transformation. The pines and oaks grow smaller, draw apart, become mere interruptions in a brown (or green, according to the season) carpet of grass that dominates the rest of the continent, either in the natural form of rangeland and prairie or the cultural form of grain field and pasture. As the Klamaths have been acting as a refuge for remnants of the ancestral forest, they have been choking the forests to the east because their rising peaks have captured more and more of the moist air from the Pacific.

It may seem farfetched of me to equate a few lush mountain meadows with the vast, dry grasslands of midcontinent. A mile or two long, Morris Meadow is one of the largest in the Klamaths. Most are hardly more than forest glades. In the Red Buttes little grassy basins of bowling-green size, once glacial bogs and lakes, now seem accidental, as though the forest simply had neglected to grow there. The Kalmiopsis has almost no mountain meadows at all, although it does have steep, dry balds of bunch grass and clover rather charitably called prairies. There are some big open spaces in the Marble Mountains, where marble bedrock is inimical to tree growth, but nothing in the Klamaths matches the extensive glacial meadows of the high Sierra.

Standing in the middle of one of these glades, though — when the heads of grasses, corn lilies, and cow parsnips block out the trees and peaks — is almost like standing in Kansas or Iowa. The plants are not quite the same, but so similar that I can't help sensing echoes. (I wonder if John Muir defended Sierra meadows from sheep graziers so fiercely because they reminded him of the lost Wisconsin prairie of his boyhood.) Swampy Klamath meadows are packed with larkspur, wild onion, crane orchids, and blue gentians, while asters, wild geraniums, sneezeweeds, and coneflowers grow in drier sites. All have close relatives native to the virgin prairie. The Klamath Mountains now may have as much virgin grassland as Kansas or Iowa, where corn and soybeans rule.

The plants are so similar that one wonders how they came to grow both on Klamath peaks and Kansas plains, whether there was an ancestral prairie from which they branched out, as with forest wildflowers, or whether they began as plants of ancestral forest glades and only later moved out into the new seas of grass. We may never know, since soft herbs fossilize less readily than woody plants. There are more plant fossils from the forested Oligocene and Miocene epochs than from the more recent Pliocene and Pleistocene.

Whatever their relationship, Klamath meadows and central prairies have the same physical origins, despite the contrast in their geographical settings. Glaciers and fire formed both. Glaciers left treeless landscapes — bogs, lakes, moraines — in both the Klamaths and the Midwest; and in both places, repeated wildfires kept fertile glacial soils from growing forest. Aspen, fir, and pine seedlings at the edge of every Klamath meadow are reminders of the meadow's fate if fire doesn't renew it by killing the encroaching trees. The trees are so persistent that one wonders how sun-loving plants managed to prevail at all before mile-thick glaciers tore up the forest. Fire must be the answer; it even seems to have left its signature on the shapes and colors of meadow plants — sunbursts of coneflower and sneezeweed, flames of paintbrush and leopard lily.

Fire and ice are, of course, classic enders of worlds. The end they make to the forest world is a very distinct one. The virgin forests of the Midwest passed into prairies just as abruptly as Klamath forests abut meadows — one moment, giant trees; the next, sunflowers that look you in the eye. Expansion, movement from forest to grassland, has been the major component of American national mythology. Manifest Destiny would not have seemed so clear if the border had stopped at the Mississippi. But then that expansion, that widening of horizons, has been generating mythologies long before ours — in the first Siberian hunters to cross the Canadian Rockies and find the bison herds, in the first nomads to penetrate the Asian steppes, perhaps in the first proto-hominids to enter the African savanna.

No landscape has more associations than grassland with the things humans have come to value most as they've spread through the colder parts of the earth: the sun as a celestial source of life, its useful earthly incarnation as fire, big herds of grazing animals. Since civilization began, successive waves of sun-worshiping, grassland nomads have dominated the human race, so that there probably is not a single system of belief without these grassland components. The human body itself seems as much an adaptation to grassland life as the hoofs of a bison. Erect gait, long legs, muscular buttocks, balancing arms, binocular vision — all are directed toward efficient movement in a landscape of horizons. If hominids hadn't left the forest, we would still be able to climb as well as apes.

If Eden often is imagined as a kind of arboretum, then heaven is seen as an equally green and flowery, but more open, place — a place where not even a fruit tree will obstruct the celestial light. Such an Elysian field has a striking likeness to an upland prairie in springtime, when herds are calving and winds smell of flowers. If a nomadic hunter or herder wanted a certain time protracted into eternity, it probably would be spring on a prairie.

There's a basic familiarity about meadows. The American pioneers may have been intimidated by the virgin prairie's vastness, but they were leaving a landscape of fields and pastures behind them as they moved west from the Atlantic. There's a comic element in the pioneers' fear that the prairie was unfit for growing crops (since it didn't grow trees) when the prairie grew big bluestem grass taller than the wheat and corn, also grasses, that they wanted to sow.

I certainly feel an expansion, an excitement, on moving out of a forest into a meadow. There is never the slight foreboding I sometimes have when entering deep woods or descending into a shady river canyon or lake basin. Meadows lack the baleful mythic connotations of lakes, rivers, and forests. There aren't any tales of travelers tricked and devoured by the demon of a meadow. Where would a goblin lurk in sunlit grass and flowers? Instead of foreboding, there is a sense of guilt at trampling a meadow in full bloom. I felt that way when camping in a meadow so remote and unvisited that I couldn't take a step without trampling some exquisite plant.

Such a meadow is a rarity; they are too attractive to go long unvisited. Few homesteads in the Klamaths are far from sunny, grassy places. (Miners' cabins are often tucked into shady gorges, but miners seek the glitter that comes from the earth, not the sky.) Meadowlands still lie at the heart of our largely tamed world, endlessly reproduced in lawns, parks, and playgrounds. How much more crucial they must have been to the old hunting and gathering world. To stand on Marble Mountain's south slope and watch deer hunters lead their horses up the green and russet parklands of Elk and Rainy valleys is to glimpse the Pleistocene world of which we are the very recent inheritors.

RUNNERS AND DIGGERS

Elk and Rainy valleys are faint echoes of the Pleistocene though. They lack the elk and grizzly bears that dominated the Pleistocene landscape, that were, in related Eurasian species, the favorite subjects of cave painters. Elk and grizzlies were extirpated from the Klamaths in the nineteenth century, and there is irony in the circumstance that these animals — so well adapted to glaciated landscape and so economically and spiritually important to prehistoric Eurasian hunters — should have been obliterated by their former ecological associates while forest-dwelling black bears and black-tailed deer, which have had much less to do with human evolution because they inhabit a continent where humans only recently have arrived, still should be thriving.

We may never know whether the first Amerindian hunters exterminated the mammoth, ground sloth, and wild horses as white settlers extirpated the elk and grizzly, but it remains a disquieting possibility. Fossils only a few tens of thousands of years old demonstrate that North American grasslands were once even richer in species than the present-day African savanna. As the black-tailed deer of today's Klamaths echoes the elk, so the elk of historical America echoed those prairie throngs. The La Brea tar pits of southern California, at most forty thousand years old, held bones of camels, bison, horses, mammoths, ground sloths, saber-toothed cats, lions, wild pigs, giant wolves, three species of condor (one with a twelve-foot wingspan), and seven species of eagles, as well as the creatures that still live in California.

One might expect to find little of grassland evolution in the relatively tiny glades of the Klamaths, but this is not the case. Unlike the red-rock forest, grassland has generated so much animal evolution that it can stock the smallest meadow with special creatures. The transition between forest and meadow animals in the Klamaths is as marked as that between trees and turf. I never get over a certain surprise when, coming to some high grassy place hemmed in by dozens of square miles of treetops, I meet the same black-tailed jackrabbits and California ground squirrels that throng oak savannas fifty miles eastward. I've seen both species right up to the peaks of the Marbles and Red Buttes, where subarctic winters must make survival a different proposition than it is for their lowland cousins. I doubt they migrate downhill through dozens of forest miles for the winter (the lifetime range of even a fleet-footed jackrabbit is only a square mile or two), so they must be resident populations. (How they arrived I don't know. Perhaps, like Jeffrey pines, they are relics of a drier period when grassland extended throughout the Klamaths.) Certainly, neither the burrowing ground squirrels nor the lanky jackrabbits are evolved for forest life. Their tawny pelts and amazing resistance to drought put them at the end of millions of grassland years.

Many meadows in the Klamaths are without ground squirrels and jackrabbits, but I doubt any are without voles and pocket gophers. Indeed, a meadow might be short lived and barren without these rodents. Voles and gophers are such abundant, tireless burrowers in meadow soil and eaters of meadow plants that they have an effect similar to a farmer's on a field — they cultivate. They aerate the soil with their diggings and fertilize it with their droppings. Their workings are everywhere in Klamath meadows: mazes of tunnels and nest holes built in the grass by voles, heaps of earth pushed up by gophers. They're so busy they pay no attention to intruders. Voles have run over my legs as I sat in the grass, and gophers have pulled underground plants against which my hand was resting.

Voles and gophers are so intimately tangled in the grass that meadows and rodents must have arrived in the Klamaths together. Voles live in herbage throughout the Northern Hemisphere, vying with the closely related mice and with shrews for the distinction of being the most abundant mammals on earth. Pocket gophers live only in North American grasslands and display a variety of strange adaptations, including the namesake pockets of skin outside their cheeks and an ability to bite roots and soil without opening their lips (which close *behind* their incisor teeth). The pockets and other features suggest that gophers are related to squirrels; if so, they are carrying to extremes the evolution of squirrels from ancestral tree dwellers to burrowers. Unlike ground squirrels, gophers seldom leave their burrows. There is no more tenacious grassland rodent, as the gopher's persistence in suburban lawns and city parks proves. Gopher bones were abundant in the La Brea tar pits.

Meadow rodents attract grassland predators. Hawks and golden eagles elicit the same alarm whistles from ground squirrel colonies in high meadows as in Sacramento Valley foothills. A red-tailed hawk I watched soaring low over Little Bald Mountain Prairie in the Kalmiopsis had to fly sideways because the "prairie" was so steep, a posture that would have complicated the business of swooping to prey. The ground squirrels perhaps realized this, since they didn't act as apprehensive as would colonies on more level ground. Or maybe their splendid isolation had made them complacent. Their world consisted of about ten acres of low bunch grass (premium ground squirrel habitat since tall grass makes it hard for them to watch for predators), surrounded by some of the biggest, densest Douglas fir forest in Oregon.

With the nearest other prairie two miles away, the squirrels may have suffered less from predators, disease, and intraspecific conflict than populations of more continuous grasslands. I don't imagine they found much to argue about with the gray squirrels that sometimes buried acorns in the prairie. If squirrels think about such things, they may have wondered at the unsquirrellike behavior of their distant cousins — jumping about in trees and eating nuts instead of digging burrows and eating grass. They perhaps wouldn't acknowledge any kinship with gray squirrels at all, like Victorian divines confronted with the chimpanzee and certain that no relative of humanity would live in trees and eat bananas instead of bread.

The only places in the Klamaths where I've heard coyotes were the high grasslands of the Marble Mountains and Red Buttes. Coyotes wander in and out of the mountains, moving from meadow to meadow in search of ground squirrels, gophers, and voles, which they specialize in stalking. An even more specialized grassland predator lives in the Klamaths. The badger is another member of the weasel family, the extraordinarily plastic group that has put a major predator in every ecological niche of the Klamaths. As the otter embodies riparian grace and the wolverine, boreal ferocity, the badger seems the quintessence of earthy strength. Almost as broad as long, it may be able to lift a thousand pounds on its back, an ability reportedly demonstrated at rodeos, where captive badgers are said to have supported a horse and rider without being crushed. A badger can smell a gopher nest through the soil and dig itself out of sight faster than an enemy can get hold of it. It virtually swims through the ground with its powerful foreclaws.

The badger is such an expert burrower that people seldom see it, which points up an aspect of grassland life that may mislead the superficial observer as the prairie's treelessness misled the pioneers. So many grassland animals live underground — not just rodents but also the burrowing owls, snakes, toads, and lizards that share their burrows — that meadow faunas may seem impoverished on the surface. But grass roots and wildflower tubers and rhizomes go deep, and more life may go on beneath a meadow's surface than above it. A meadow may have just as much, or more, living biomass as a forest, most of which is comprised of nonliving wood.

The dryness and cold that forced grassland evolution also forced grassland organisms to demonstrate greater efficiency than is typical of forests. Grassland packs more energy into smaller spaces than forest. It reproduces with grains and thistle seeds instead of acorns and pine nuts. Consolidation of plant energy increases animal density, either in the obvious form of huge, grazing herds or in the hidden but even denser burrowing faunas. "Gopher" derives from the French term *gaufre de miel*, honeycomb. I've seen glades in the Klamaths where there were more gopher diggings than plants, so that I wondered what the gophers found to eat. The dry ground in such places gives way swampily under one's feet.

Meadow wildflowers are extraordinarily abundant producers of food energy, as the numbers of nectar and pollen feeders thronging them show. It's hard to imagine how more vertebrate energy might be stuffed into a smaller package than the hummingbird, without reaching some new biotic plateau, perhaps an organism so energetic as to be invisible to our slower senses. I remember watching a calliope hummingbird, the smallest United States hummingbird, on the Salmon Crest. Hardly larger than a bumblebee, it seemed to appear and disappear on a branch until I'd stared at it long enough to ascertain that it was flying down to feed in a patch of paintbrush from time to time. The change from perching to flight simply had been too quick for the casual eye to perceive.

Birds have populated grassland almost as effectively as mammals. Some early grasslands were dominated by birds, as the giant moa bones of New Zealand and the legendary rocs of Madagascar imply. Such enormously powerful, flightless birds make a dinosaur ancestry for them seem quite probable. Mammals won the competition for the gigantic, earth-treading grassland niche, but in smaller form birds are just as common as mammals in Klamath meadows. The finches and sparrows, the most recently evolved of all birds, are particularly dependent on grassland since they eat mainly seeds. I've seen white-crowned sparrows plucking and eating grass seedlings, unusual behavior for birds. The bird's beak and crop is better adapted for crushing high-energy seeds than for extracting sugars and proteins from the tough cellulose of grass plants. Even the mammalian stomach wouldn't be equal to that task without the symbiotic help of cellulose-digesting bacteria.

Rich and progressive as they are, grasslands still place limits on their inhabitants. Young worlds contain dangers as well as opportunities. The vanished Pleistocene herds are evidence of that. Huge and extravagant, they suggest the instability of experiments, melting away before a change of climate or a spear-wielding primate, while more inconspicuous, burrowing life, closer to the grass roots, goes on with serene conservatism. Rock strata of the past twenty million years hold an array of titanic but curiously evanescent creatures — pigs with four-foot-long skulls, four-tusked mastodons, giant camels, dog bears — while rabbits are shown to have remained much the same.

We shouldn't take grasslands for granted, comfortable though we may feel in them. There's a spectral aspect to our Elysian visions. Seen in the distance, across midwestern grain fields, the high-rises of our corn-fed cities can look like mirages, ready to be swept from the horizon by the first change of wind or sudden storm. They risk the vulnerable gigantism that left nothing of titanotheres except fourteen-foot-tall skeletons. Young as it is, grassland is not exempted from the old circular imperatives of life that require organisms to go down into darkness as well as up into light. If one looks carefully (and one had better), one will find more than hummingbirds and rabbits in a meadow.

SNAKES IN THE GRASS

At Morris Meadow I walked into a particularly lush growth of grass and herbs to get water from the creek, and it seemed that a big black or brown snake slithered before me at every step I took. I found more snakes in that meadow than I did in three days of canoeing across the Okefenokee Swamp. (I found no snakes in the Okefenokee, but then I was there in November.) Inappropriate as a serpent may seem to something as heavenly as a mountain meadow, it fits in quite nicely, which isn't surprising. If snakes evolved by retreating to an underground, burrowing life then returning to the surface transformed, grassland was a good habitat for such transformed creatures to occupy. Even though snakes evolved long before the great grasslands of recent epochs, as fossils prove, they owe much of their present abundance and diversity to grasslands and their teeming rodent and bird populations.

Some snakes, such as rattlesnakes, seem definitely to have appeared during the spread of the prairies. The pit-viper family, to which rattlesnakes belong, is one of the most modern. Rattling may

have evolved as a device to prevent trampling of the snake by warning hoofed mammals of its presence in the grass. The rattlesnake's
venom and its heat-sensing pits are adaptations for preying on small
ground mammals such as voles and gophers.

There certainly is no more efficient way for a small reptile to
make its way through grass than the snake's sinuous crawl, as anyone who has tried to catch a snake in a meadow will agree. "The
snake in the grass" is more than a cliche, it is a piece of true folk
ecology, and it evokes echoes of early myths in which the snake was a
god. The vipers, cobras, and mambas of African savanna comprise
one of the greatest arrays of virulently poisonous snakes on the
planet. This could have had a profound effect on the developing
human imagination as proto-hominids evolved out of an original
forest environment, in which their ability to climb trees and focus
on the forest floor had protected them, into grassland, where they
often couldn't see where they were putting their feet.

Did sudden snakebites reinforce anthropoid stirrings of uneasiness over the clubs and throwing sticks, the bones of fellow primates
and antelope cracked for marrow? Like lightning, another danger
of open horizons, did the snakebites seem a kind of retribution, a
taking back for what had been given? Or later, when the much more
efficient Paleolithic hunters swarmed out across the Eurasian
steppes, was there foreboding as the giant herds melted away, year
after year, species that could feed a family for months disappearing
within speaking memory, too many families fed? Myths of virgins
sacrificed to dragons are so widespread as to suggest a deep-rooted
notion of serpents as arbiters as well as symbols of fertility.

Nobody knows when blood guilt began — it has left no fossils —
but all myths evince an uneasiness at the necessity of killing to live,
an uneasiness that goes back to the beginning of things. In classical
myths the world begins when the Olympian gods kill their Titan
parents. In even older, tropical myths a hero kills the earth's consort
snake god, from whose dismembered and buried body yams and
breadfruits grow. The snake god is the generative force that allows
itself to be cut up and eaten, but what has been taken must be restored, or the herds or fruits may not return. In the Eleusinian fertility rites of ancient Greece, live pigs and human figures made of
bread were thrown into underground chambers that were inhabited
by snakes. Shamanist and animist societies thought that all deaths

were caused by spiritual forces, and that animals possessed spirits just as humans did. Snakebite must have seemed a deliberate, highly significant event to them.

It's easy to understand how prehistoric people might have had a preoccupation with snakes. I have one myself when walking in ratttlesnake country, especially in grassy places. It's more than the simple caution I feel about falls or other accidents (which cause many more wilderness injuries and deaths than snakes do). All kinds of irrational, and thus probably very deep-seated, feelings creep in. I have a nervous fantasy of snakes crawling into my sleeping bag. Everybody who's slept out probably has had this fantasy: it must be archetypical. Maybe we vicariously experience some rite of passage or trial by endurance through it. One of the things that makes wilderness memorable is the possibility of facing death so closely, a possibility that can greatly enhance one's commitment to life.

I also can understand how prehistoric people thought that snakes communicated with them. I've had such thoughts. Several times I've been walking heedlessly along a trail, suddenly felt an impulse to be cautious, and proceeded carefully for a few yards to find a rattlesnake quietly sprawled or coiled in my path. On a grassy trail in the Red Buttes my companion asked why I was walking so slowly, and, a few seconds after I told her I was looking for rattlesnakes, we found one in the herbage beside the trail. I felt quite grateful to that snake — seldom are one's preconceptions confirmed so distinctly. The only rattlesnake I've encountered when not feeling cautious was the one that rattled at me along Big French Creek. It also was the only one that has ever rattled at me, as though it had fallen back on a cruder warning after a failure of telepathy.

Prehistoric people, living in wilderness, must have had such experiences with snakes frequently. It's easy to see how they viewed them as messengers from a buried god of fertility. People still are fascinated with snake holes, as though they are not simply burrows from or into which snakes are seen to glide, but openings to a dark realm where the god still lies, supporting with his body the tree of life, as does the primal giant Ymir of Norse mythology.

Grendel and his lake full of serpents comes to mind here, or, more recently, King Kong in his cinematic battle with what looked to me like a giant viper. Associations of mythic wild men and giants with serpents are legion. Sometimes they are allies, sometimes enemies, but the association generally refers to some interpenetration of human origins and serpents. Typhon, the Titan killed by Zeus in establishing the Olympian hegemony, was half giant man, half giant snake.

It is easier to discern this interpenetration in mythology than to figure out what it means. Assuming that uneasiness about shedding blood is a peculiarly human trait, though, I wonder if Pliocene visitations by snakes in the grass might have had a formative effect on the idea of justice. In classical mythology blood guilt was punished by the Furies, daughters of a wounded Titan, attendants of Proserpina (queen of the underworld), wearers of serpents in their hair. Perhaps the rule of law began in African savanna, when the buried god sent snakes to even the score with unprecedentedly clever killers of antelope and baboons. Surviving a snakebite certainly would put the fear of the gods into me.

How would Klamath giants feel about snakes: would they be as uneasy about them as we are? Reports indicate they don't use weapons and are only incidental predators, so maybe they wouldn't share the blood guilt that put our ancestors in awe of snakes. In this case, lacking the imagination to feel guilty, the giants might not be what we consider human. Blood guilt is not evident among beasts. The knowing slaughter that has given us an uneasy sense of our own mortality, our own complicity in the cycle of life and death, was a necessary precursor to justice, mercy, and other humane concepts.

But perhaps I'm just being anthropomorphic in equating humanity with blood guilt. I suppose a completely vegetarian species could develop imagination and the empathies and insights that go with it. Would that species begin to feel guilty about killing plants? The uncivilized mind sees spirits in everything, not just in animals. An association between the realization of one's own mortality and one's mortal effects on the living things one must eat would seem an inescapable act of dawning imagination, whatever the species.

Grassland may be the youngest landscape on earth, but for humans, in a sense, it is the oldest. Not only has grassland shaped our bodies, but much of our minds have been shaped by it too, as indicated by its mythic resonances. It is no wonder we feel drawn to wide-horizoned landscapes like the ocean, alien though the ocean may be to our immediate evolutionary background. We are used to having our eyes on horizons, and where we do not find them, we have learned to make them, pushing aside whatever obstructed our view of our prospects. Even more than fire and ice, we have become the great spreaders of grassland on the planet. Civilization is a grassland symbiont, fully dependent on the condensed food energy of grain.

If waterways are like youth and forests like maturity, then grasslands are like old age. There is a similar apparent shrinking of the body, the forest masses of trunks and branches dwindling to stems, as human muscles dwindle toward the bone. A meadow in winter is an emblem of the sere, without even a bare bough. But there still is vitality beneath the wrinkles and baldness: in humans, a life's accumulation of knowledge and experience; in grasslands, something similar, an accumulation of evolutionary adaptations (knowledge and experience of a sort) as old as the planet.

Grasslands also are like human old age in that they begin to take on aspects of infancy. Many Klamath meadows grow around, or in former beds of, glacial lakes. Thus they are connected to all the teeming primitiveness of lakes, with their crustaceans and mayflies that might as well be living in the Paleozoic era. At a glance the first terrestrial landscapes must have looked like grasslands, patches of small green plants in moist soil. The childlike innocence one feels in a meadow might even be compared to human senility — a fading away from the complexities of maturity into an oceanic security and simplicity of faraway memory, of the world's beginning, when all life waited, without guilt or anxiety, to be lived. In a meadow, evolution seems to grasp its own tail in its mouth, like the mythical snake that simultaneously devours and gives birth to itself.

There is a limit to this analogy, as to all analogies. A human life ends with old age (on earth, at least) and there is no indication that evolution means to end with grasslands (though civilization may be doing its best to accomplish this). My convenient symmetry of a Klamath Mountain lake ending in a Klamath Mountain meadow is biologically flawed because meadows are not the final fate of glacial lakes in the Klamaths; forests are. Unless fire or renewed glaciation prevents it, the forest eventually will reclaim the rich soils left by silted-up lakes that meadows now occupy. The fact that grassland evolved after primal ooze and forests doesn't necessarily mean that it will outlast them. The climate could change back to what it was in the Cretaceous period. Then the great temperate forests of sixty

million years ago would form again, burying our grain-based mega-civilization in a tangle of tree roots and rain-carved, rocky hollows —
a continent-wide Appalachia. Or entirely new climatic and geological conditions might produce an unprecedented landscape, forcing
grassland into protected pockets, as the ancestral temperate forest
has been forced.

Compared to forest or aquatic ecosystems, grassland is unstable.
It requires rather precise geological and climatic conditions, and if
these conditions are not maintained — if too much rain falls, or too
little — it quickly turns into forest or desert, both of which are domi-nated by woody plants. This instability is reflected in the spectac-ular but brief careers of various grassland faunas. Humanity, with
its dazzling symbioses, preadaptations, and neoteny, is the most
spectacular of these, and may well be the briefest. The grassland
soils of North America, the richest in the world, are in a state of
hemorrhage, bleeding away into creeks and rivers by the millions of
tons each year as modern agriculture gouges crop after heavy crop
from the prairie. When the soil is gone, the high-rises will go too. Of
course, this won't be the end of the world; evolution proceeds
through impoverishments as much as through enrichments, and
there will be something after the grain combines have cut their last
row. Some plant that uses energy even more efficiently than grasses
may be evolving now in herbicide-laced drainage ditches.

Evolution probably will end sometime, when life ends. But it is a
much less predetermined phenomenon than human life, for all the
parallels that can be drawn between the two. This is perhaps how
evolution as a mythology differs most from preceding myths. None
are as open ended as evolution. None look upon the future as simply
a gigantic question mark. Whatever they have foreseen — heaven,
hell, *Götterdämmerung*, endless reincarnations, worker's paradise,
galactic empire — other mythologies have envisioned the future as a
fulfillment of a predetermined plan or process, a fulfillment where-in the end somehow recapitulates the beginning, as old age recapit-ulates infancy. Original sin is resolved in salvation or damnation.
Enlightment is regained as the serpent dream of mortality ends. The
social equality of the tribe is reestablished in the industrial state.
The acquisitive restlessness of grassland nomads is transferred to
rocket ships for galactic exploration.

There may be symmetries inherent in evolution. Certain themes

—plant and animal, producer and consumer, reproduction and death — may be carried through from beginning to end. Perhaps symbiosis will continue to produce unprecedented new life forms from cooperative associations of old ones. But we have no way of knowing. The lesson of dinosaurs and early mammals is too clear; there quite likely never would have been an Age of Mammals if dinosaurs hadn't disappeared first. There are no chosen inheritors.

Such open-endedness can be frightening, but it also can be liberating. It implies a freedom for life that is absent from the oldest mythologies, with their notion that things will remain the same until the end of this spirit incarnation, and from various newer ones, with their notion that life will henceforth be controlled by our present values. Both notions seem static to me, seeing no future but the one desired, a perfection of present values; or the one feared, a loss of them.

I said earlier that evolution as a myth has lacked a clear ethical dimension. Certainly, there isn't much moral instruction in the dinosaurs' rise and fall. I wonder, though, if the very lack of pre-determination that can make evolution seem amoral may have an ethical potential — at least, in an evolving world. Ethical questions become qualitatively different raised in a five-billion-year-old world than in a five-thousand-year-old one. They're stretched out of shape by enormous time spans. They fall prey to evolutionary relativism even now, as we try to determine how many people can live with justice and mercy on an earth that is not going to get bigger anymore. Justice and mercy for the living may be a cruel swindle for their overpopulated and impoverished descendants.

In a world where such ambiguities reign, the idea that there can be no predetermined future may be a salutary one. It allows for flexibility, and nothing is more necessary in dealing with the ambiguous. One might say, indeed, that evolution itself has been a history of life's flexibility in threading the ambiguities of environment. It is the flexibility of the grass plant that has allowed it to prevail between heat and cold, fire and ice, wet and dry. The farther an organism evolves away from flexibility, the less likely it is to survive change, which is inevitable. The contemplation of evolutionary question marks, serpent shapes poised above planetlike spheres, may be as instructive as the contemplation of icons, stars, or navels.

CHAPTER NINE

THE HUMAN ELEMENT

There were giants in the earth in those days;
and also after that.

Genesis 6:4

HUMANITY HAS ALWAYS been hard to define, and evolution hasn't made it easier. Older myths generally placed humans on a scale midway between animals and gods. This position had a comfortable stability. It gave people something to look down on and something to look up to. Some evolutionary myths have repeated this formula, with the idea that humans, having evolved from animals, will presently evolve into superintelligent beings somewhat like the gods of earlier myth. This is understandably the most popular kind of evolutionary myth. It takes dozens of forms, from Teilhard de Chardin's noosphere to Nazi superman eugenics.

A godlike future for the human race may be possible, indeed desirable, assuming our future godlike omnipotence and immortality are accompanied by better behavior than that of, say, the Olympian gods. Evolution's four billion years on this planet do not foreshadow such a future, however. The symmetry of transformation from animal to god is not reflected in evolutionary evidence. Humans have not evolved from animals; we *are* animals, no less dependent on plant photosynthesis and bacterial decomposition for our survival than the lowliest flatworm. The ancient thinkers who developed the animal-human-god hierarchy were not aware of what we *have* evolved from. Like all animals, we have evolved from an intricate, fortuitous symbiosis of single-celled organisms. If there *is* symmetry to evolution, the future will not see us dominating all other life as gods. It will see us become part of a greater organism which we cannot imagine.

Evolutionary humanity is a truer microcosm of nature than medieval philosophers dreamed. The human body does not merely resemble nature in its parts, it recapitulates the history of life, as much as living reenactment of evolutionary dramas as the Klamath Mountains. Corpuscles float in a primal nutrient bath of blood; intestines crawl about absorbing food in the manner of primitive worms; lungs absorb and excrete gases as do gills and leaves. No human organ would look out of place if planted in some Paleozoic sponge bed or coral reef. Even our brain is an evolutionary onion, the core we share with fish and reptiles, the secondary layer we share with other mammals, and the outer layer we share with other primates.

Humanity can't be defined apart from the intricacies of natural selection, mutation, symbiosis, preadaptation, and neoteny that formed it. It can't be defined apart from the millions of other species on the earth. Our evolutionary myths have been greatly oversimplified in our attempts to follow the thread of humanity into the past (and the future) while we ignore its entanglements with the threads of grasses, trees, snakes, and other beings. Such myths make a falsely passive background of an actively evolving world. To say that humanity descended from the trees, adopted a grassland hunting life, then invented agriculture and civilization, is racial solipsism. It would be quite as accurate to say that the forest abandoned the hominids, that the grasslands adopted them, that the first domestic plants and animals chose to live with our Neolithic ancestors. I could write an evolutionary history of the human species in which its main significance is not as an inventor of language or builder of cities but as an ally of grasslands in their thirty-million-year struggle with forests. An extraterrestrial observer of the human colonization of North America would have seen more of spreading grasslands than of spreading cities, grasslands spread first by Indians with fire, then by whites with axes and plows.

It is possible to look upon humans and their civilization as a biological and geological force not qualitatively different from the volcanic eruptions, glaciations, and other catastrophes that have disturbed organic evolution. Nuclear war and wholesale industrial pollution may do life on earth more damage than a billion years of exploding volcanoes, but anthropoid greed and convection currents in the earth's mantle seem about equally random and senseless. Molecules simmering in the skull of a primate or sixty miles underground — what's the difference? Both explode when pressures get critical.

Such a view falls into the error of seeing evolution as a predetermined phenomenon, though. A humanity destined for demonic holocaust by its manipulative cleverness is a mirror image of the more popular evolutionary myth of a humanity destined for godlike triumph. If the four billion years of evolution demonstrate one thing, it is that humanity is not *destined* for anything. Evolution has always been open to new possibilities, which is why it has been so chaotic and devious. Every organism continually confronts a galaxy of evolutionary choices.

The difference between humans and other organisms is that humans, having discerned something of how evolution works, are now able to confront their choices consciously. This is not the same as saying that we now can *control* evolution. I don't know how much of a difference it is in effect: we may be able to perceive our choices and still be unable to choose and act. By overpopulating the planet as we are now doing, for example, we are making an evolutionary choice just as unplanned as that of our hominid ancestors when they began cracking antelope and other hominids over the head with sticks. Nevertheless, we do differ from the first hominids in our having some notion of the implications of our behavior. In Biblical terms we have heeded the serpent, eaten of the tree of knowledge, and lost our innocence. We now must face the possibility of choosing between good and evil, or, in evolutionary terms, between survival and extinction.

In other words humans have some degree of free will. As two millennia of theologians have been telling us, this is a perilous position. Pride is the great danger to the soul consciously seeking salvation. I think it is the great danger to the species consciously seeking survival too. In both cases pride can transform the best of virtues into the worst of vices. It can transform an individual's high intelligence into arrogance, and it can transform a species' considerable understanding of nature into stupid plundering.

The King of Phrygia tied the Gordian Knot in the temple of Apollo and prophesied that whoever untied it would become Lord of Asia. Alexander, proud young conqueror, cleverly cut the knot with his sword, became Lord of Asia, and died at age thirty-three of alcoholism, disease, or poison. Apollo was a god of the serpent as well as of the lyre. Conquering civilization could cut the Klamath knot, and that of every other wilderness: dam every river, log every forest, plow every meadow, until the last gasp of splendor subsides from the earth. "What now?" the serpent might whisper, as it perhaps whispered to Alexander on the banks of the Ganges.

The mythic resonance that evolution has given the natural world expresses a multitude of choices that humans have not consciously faced before. In old myths wherein nature remained the same from the world's creation until its end, our relationship with nature was much less laden with choices. Men couldn't change what the gods had made: "saving the planet" would have seemed an impertinence. But evolution, wherein a bear is not simply a black, shaggy animal but a wave of animals surging up through abysses of time from the original one-celled beings, raises troubling questions. Should we follow its competitive trend, manifested in natural selection, and try to survive by destroying everything that seems to get in our way? Should we follow its cooperative trend, manifested in symbiosis, and try to coexist with our parasites and our hosts (whoever *they* may be) in hope of some new synthesis? Should we try to do both? After all, that is what evolution does.

Much of the disquiet that has beset our thinking in the past two centuries seems related to evolution's burden of new choices. If life has taken such different shapes in the past, who can feel any assurance about the future? Entire realms of confident human activity begin to seem absurd. This is not necessarily a bad thing, of course. Some of the worst atrocities have been committed in pursuit of assurance, in flight from anxiety. If the future is essentially unknowable, at least good ends can no longer justify evil means. If it dispells our dreams of heaven, a world without destiny also wakens us from nightmares of hell.

The giants who left their tracks near Bluff Creek are eloquent mythic expressions of evolutionary uncertainty. Are they competitive lords of the snow forest? Cooperative children of the ancestral forest? Are they human? Are they alive? In a sense the giants are the missing link that Victorian society demanded Darwin and Huxley produce before it would bow to the new version of genesis. (Newspaper articles of 1884 tell of a young giant captured in British Columbia and shipped to London alive. It never arrived, victim, perhaps, of some conspiracy of Anglican divines?) Giants express our familial relationship to the rest of life. If we found them, could we rightfully continue to clear cut the giants' forests, dam their rivers, and trample their meadows? Even giant-hunters who consider them "just animals" advocate creating large preserves for giants.

With their elusiveness to civilized knowledge, giants express a gap that has arisen between our thinking habits, which are expressed in everyday speech, and our very recent awareness of evolutionary evidence. We condemn "brutality" and scorn murderers as "animals." We fear the sight of a shaggy beast shaped like a human. Yet we know that no wild animal is remotely capable of the deliberate torture and mass extermination that have become common in this most civilized of centuries. Knowing how unprecedented these horrors are, we no longer can blame them on our "lower" animal instincts, or hope to escape them by "rising above" our animal nature. We will not rise above our animal nature until we begin to live without food, water, and air. We are more protected by the timidity of the wild animal that remains in us than we are threatened by its aggressiveness.

We are fortunate to have the self-consciousness that allows us the possibility of free will. But we no longer can assume that our consciousness imbues us with a predetermined destiny separate from the rest of life. The only way we can separate ourselves from our animal, plant, and fungus relatives is to stop living, a viable and popular evolutionary option (considering the millions of extinct species) but one we're self-consciously averse to. Yet as we develop from species-exterminating hunters to land-eroding farmers to biosphere-polluting industrialists, we increasingly separate ourselves.

Few organisms survive in rapidly changing environments, and the world is changing faster than ever before. The fact that we've set these changes in motion doesn't mean we can control them. *We* must change to survive. No biological change will be fast enough now, though; we can't evolve as fast as the insects or rodents or microorganisms we've "conquered" because we reproduce so much more slowly. We must depend on cultural evolution. If our behavior is to change, our myths will have to change.

Myths began as imaginative projections of human consciousness onto nature. Trees had language, birds had thoughts, spiders had technology. When science found that nature does not, in fact, have a human consciousness, some thinkers concluded that myth was dead, that there was no further need for imaginative views of a world which, they thought, had no consciousness at all. But they misunderstood science. That nonhuman life has no human consciousness doesn't mean it has no consciousness. Science has opened a potential for imaginative interpretation of nature that is enormously greater than the simple projection of human thoughts and feelings onto the nonhuman. It has allowed us to begin to imagine states of consciousness quite different from our own. We can begin to see trees, birds, and spiders not as masks concealing humanlike spirits but as beings in their own right, beings that are infinitely more mysterious and wonderful than the nymphs and sprites of the old myths.

Science has raised the possibility that there are as many different consciousnesses in the world as there are organisms capable of perception. It also has raised the possibility that consciousness may arise in ways that seem very alien to us. The symbiotic superconsciousness I vaguely sense in forests is not outside scientific possibility.

The age of myth is not dead; it is just beginning, if humans can survive to inhabit it. Only, instead of myths peopled with talking trees, we must begin to create the opposite. (The fact that such myths — inhabited by "treeing talks" — aren't fully expressible with our present syntax and vocabulary is one measure of the magnitude of the enterprise.) Instead of inflating our human consciousness to fill trees, we must let the trees into our minds. It is not a sentimental undertaking. When science found that we don't have thoughts and feelings in common with the nonhuman, it also found we do have something equally important in common — origins. We are very different from trees, but we also are like them. As we learn how they live, we learn a great deal of how *we* live.

Learning does not occur only in the mind. High towers of intellectual learning require deep foundations of emotional knowledge, or they lack stability. The more we know about trees, the more we need to feel about them. The human element has grown too large and powerful for petty or trivial feelings about the nonhuman. What we feel about pettily, we begin to destroy, as we are destroying forests to produce junk mail and other trivialities.

Future myths will be different from past myths, but their function will be the same — to sustain life. When the human element was small, when there were billions of trees and only thousands of people, it was sustaining to imagine that trees contained spirits humans could talk to, propitiate, befriend. It gave proportion to the world. Now, when there are billions of people, and not so many trees, it is sustaining to imagine what it might be like to open one's flowers on a spring afternoon, or to stand silently, making food out of sunlight, for a thousand years. It gives proportion to the world.

Of course, imagination can only go so far. The incompleteness of scientific knowledge also limits emotional knowledge. We can't fully imagine a tree's existence because we don't know how, or if, a tree experiences its life. So something of the old mythological imagination probably will linger for a long time. We will continue to project our human feelings onto other organisms, as we try to imagine their nonhuman experience.

As with organisms, new myths don't appear fully formed, but evolve imprecisely out of old myths. Giants may be an example of such evolution. Giants seem to have originated as a way of giving human form to all that is titanic and inchoate in nature. In human form the awesomeness of rocks, waters, and tangled vegetation could be wrestled into submission, even befriended, by heroes and gods. Today's Klamath giants have something of this. In their dominance of the awesome snow forest, the giants affirm a desire for human power over wilderness, for a linkage with nature that is advantageous, albeit peaceable. If we found the Klamath giants, we would grasp some essence of the titanic knot of rocks, waters, and trees, as Beowulf and Gilgamesh grasped their ancient lands by defeating Grendel and Enkidu.

But the Klamath giants also have become more than shaggy, beetle-browed projections of human desire. We begin to see in them the possibility of a consciousness quite different from our own, of a being that may be very close to us in hominid origins, but that may have evolved in mysterious ways. We imagine an animal that somehow has understood the world more deeply than we have, and that thus inhabits it more comfortably and freely, while eluding our self-involved attempts to capture it.

Giants might be seen as a kind of preadapted myth that can help us to survive the world we've created. Giants have hovered for thousands of years in the backgrounds of our dreams of immortality and omniscience, large shadows humans cast behind them as they moved toward brilliant visions of limitless power. But now the visions are fading into a natural world that has proved much deeper than we ever had imagined. Giants can have a new function in an evolutionary myth. They link us to lakes, rivers, forests, and meadows that are our home as well as theirs. They lure us into the wilderness, as they lured me, not to devour us but to remind us where we are, on a living planet. If giants do not exist, to paraphase Voltaire, it is necessary to invent them.

The Klamath Knot is one of the lucky books that came at the right time. American conservation has had a blind spot for the complexity and diversity of the Klamath Mountains. John Muir bypassed their shadowy gorges on his trajectory from the Sierra Nevada's shining grandeur to Alaska's, and attitudes toward western wilderness largely followed him. By 1983, however, the Klamaths had a following of naturalists and conservationists who loved the region and hated the exploitative management its obscurity has allowed. Their work (John Hart's 1975 *Hiking the Bigfoot Country*, for example) helped me write the book, and their support helped keep it in print for two decades. One grassroots activist, Lou Gold, said *The Klamath Knot* "described" the region for people like him. The Siskiyou Regional Education Project which he helped found gave away hundreds of copies as a membership premium.

The Klamaths are still relatively obscure. When I mentioned them to a *San Francisco Chronicle* environmental reporter in the 1990s, he said: "Where's that?" So the book remains timely. But reissuing it raises a problem. *The Klamath Knot* is not strictly a natural history guide, but a personal essay on the region's ecosystems, seen through the lens of what I call "evolutionary mythology." As such, it wouldn't lend itself readily to the piecemeal corrections of a guide revision, although inaccuracies exist. So I haven't revised it, and there may be more to this than preserving my prose, because the various kinds of inaccuracies show how the region and our understanding of it are changing. I'll point them out and correct them here, as far as I can.

First, the book left out some of the Klamaths' interesting places, an inaccuracy that arose partly from logistics, partly from the sloth Ronald Reagan endorsed when he asked: "How many trees do you need to see?" When I backpacked up the Illinois River trail in 1979, for example, I didn't reach a spot called Bald Mountain because someone raided my food, and because I thought it would be like the other ridgeline prairies I described in the "High Meadows" chapter. When I went there in 1988, I found a very different place, a giant's park of enormous Douglas firs in a sward of grasses and wildflowers. I also found an object lesson in the ecological role of wildfire. Big fires had hit the Klamaths in 1987, charring the trees at Bald Mountain, but most had survived. It was clear what had made the giant's park — "Demon Fire," as the Forest Service called it in the 1980s.

I usually found surprises when I visited places I'd missed. The most spectacular surprise was at Devil's Punchbowl in the High Siskiyous, a site known in botanical circles as one of the southernmost occurrences of Alaska yellow cedar. I meant to visit it during my 1979 walk up Clear Creek, but the illness I described in "Tracks in the Wilderness" intervened. Anyway, I told myself, it was probably just like the other glacial cirques in the Klamaths, a gentle basin covered in red fir and mountain hemlock, with a few big Alaska yellow cedars thrown in. When I went in 1992, I was glad I hadn't tried to get there after my rough night thirteen years earlier. The climb is grueling, uphill for miles, and the top is anything but a gentle basin, a savage gorge where thousand-foot granite cliffs brood over a raw black tarn, and sparse forest straggles on the scree. The glacier that carved it must have been very deep. Instead of the large Alaska yellow cedars I'd expected, I found a dense shrubbery of stunted specimens, with few standing over six feet, although the species grows to one hundred feet tall further north. I camped in a "grove" of twelve footers that looked ancient. The trees clearly were at the limit of their adaptability, although the Punchbowl seemed "Alaskan" enough.

My biggest oversight was Russian Peak, famous for its world record of seventeen conifer species within a square mile. I'm not sure why I missed it before 1983. Perhaps I assumed it would be another typical, gentle cirque. Indeed, as I found in 1987, it is a complex of typical, gentle cirques cradled in a knot of granite peaks. But it is far inland from the lush Siskiyous, surrounded by a sea of ponderosa pine, so I was impressed when I climbed above 4,000 feet to see a forest equally lush, a mossy wall of incense cedar, Douglas fir, white fir, red fir, Jeffrey pine, sugar pine, and western white pine, with here and there the jigsaw bark and weeping branchlets of Brewer's spruce. An hour's walk from the trailhead, I was more impressed to find big Englemann spruces lining a streambed over an understory of western yews. Seeing the blueish-needled spruces, which occur in only one other place in California, was like being teleported from the Siskiyous to Yellowstone.

The trees articulated something tricky about the Klamaths. Because of its evolutionary surprises, the region has been called a "temperate coniferous Galápagos," but this is a misnomer as it applies to trees. The Galápagos' marine iguanas and other unique species have evolved fairly recently to adapt to new habitats, which is why Darwin made good use of them. They are "neo-endemics," which embody evolutionary change. The Klamaths' Brewer's spruces, Port Orford cedars, and Sadler's oaks are "paleo-endemics" which have stayed the same to remain adapted to remnants of old habitats. Unlike the region's many neo-endemic wildflower species, some of which evolved in thousands of years, the trees have changed little over millions. They embody evolutionary stability, which we understand less than change. Like many mammals, humans tend to evolve quickly, and it is easier for us to think about a new species than an old one. The Klamaths have more to teach about this than the Galápagos.

A second kind of inaccuracy in the original text is less edifying — mistakes. When I wrote in "Primal Ooze" that deep caverns might underlie the Marble Mountain Wilderness, exploration already had revealed one of the largest cave systems west of the Mississippi there, as an explorer later told me. Hiking to Marble Valley again in 1999, I was surprised at how blind I had been to cave phenomena along the trail — sinkholes and streams flowing from underground. I've heard that the Marble Mountain caves are too cold to have as diverse a fauna as eastern ones like Mammoth Cave, although the smaller system in Oregon Caves National Monument has a rich fauna, including a cricket relative, a grylloblattid, which usually lives on ice fields and may have moved underground as climate warmed after the last glaciation. The Marble Mountain caves are still little known, however, and may have surprises in store. Oregon Caves yielded fossil jaguar bones recently.

When I wrote in "The Evergreen Forest" that lungless salamanders live mainly in North American temperate forest, herpetologists already had found over a hundred species in Central and South America, as I learned when one showed me some he'd just brought back. The tropical species are even more forest-adapted than temperate ones, living inside fallen logs rather than under them, and in the moss and bromeliads of the forest canopy. When I wrote in the same chapter that the most primitive living rodent, the aplodontia, has to eat soft vegetation because of its teeth, I hadn't looked at aplodontia teeth. They are formidable (albeit less formidable than many other rodents'), as I learned when a paleontologist sent me a picture of a skull. It is their unspecialized kidneys which unfit aplodontias for the dry or toxic vegetation more recent rodent groups can eat.

For some reason, readers didn't notice my mistakes as much as a piece of awkward writing in "The Evergreen Forest" where I listed common Klamath forest plants in a passage about the Kalmiopsis Wilderness's Chetko River Gorge. I meant to evoke the forest's overall diversity, not to imply that all the plants occur in the gorge. Plants like redbud and buckeye don't. Some readers thought that was my implication, however, and one phoned just to ask if I'd really seen all those plants in the gorge. I think he hoped I'd say yes.

A third kind of inaccuracy, less embarrassing to me, has arisen from changes in our knowledge about evolution and the Klamaths. The most striking is in "Primal Ooze," where I repeated the then-accepted notion that Paleozoic lobe-finned fish evolved into legged amphibians after they began spending part of their lives on land. Paleontologists since have proved that lobe-finned fish evolved into animals with legs, now called tetrapods, *before* they left the water. Early tetrapods' legs were too sprawling to do anything but guide their owners over the bottoms of streams and swamps. I wish I had known this when I was writing the book, because, even better than the old version, it supports my point that the vertebrates' move to land was a side-effect of their adaptation to water.

Another scientific change is less fundamental, but interferes more with a point I was trying to make. I wrote in "Running Water" that steelheads and their freshwater counterparts, rainbow trout, were more closely related to Atlantic than to Pacific salmon because in 1983 their genus name was *Salmo*, the same as Atlantic salmon. This synonymy made me wonder how a seagoing Atlantic genus had crossed a continent to evolve into a seagoing Pacific species, and I found it a poignant example of the epic journeys that salmonids make to perpetuate their anadromous existence. Fish taxonomy dealt summarily with poignancy in this case, however. In 1989, it cut my knot of steelhead evolution by changing the genus name of steelheads and rainbows from *Salmo* to *Oncorhynchus,* the same as Pacific salmon like chinook and coho. Steelheads and their ancestors didn't need to cross a continent to get into North America's Pacific drainage.

Of course, changing the name still doesn't explain how freshwater rainbows evolved into anadromous steelheads or vice versa (As before, the two are classed as a single species, now *Oncorhynchus mykiss*.) The Klamaths may hold a clue to the relationship, however. Steelheads there differ from others in that they commonly make a "pre-spawning migration" back to streams after one summer in the ocean instead of remaining in salt water throughout the two or more years which usually precede the spawning run. They may do so because, like Brewer's spruce and Port Orford cedar, they are relics of an older time. Klamath Mountain steelheads may be relics of an ancient population which was less adapted to salt water than today's, and which thus returned earlier to natal streams.

Other scientific changes have affected my text more subtly. When I characterized ideas of symbiotic cell evolution as "pretty unscientific" in "Primal Ooze," I was being overcautious. Such ideas are now widely accepted. When I characterized soil organisms as central to Klamath ecosystems in "The Evergreen Forest," I didn't know just how central they are. Studies publicized in the 1980s show that a cubic yard of ancient forest soil may contain hundreds of arthropod and other invertebrate species and thousands of fungal, bacterial, and protozoan species, all vital to the maintenance of five-hundred-year-old trees. When I characterized Klamath Mountain rocks as lively, even prankish, in "Rock Bottom," I didn't know how lively some may be. "Suspect terranes," chunks of land rafted eastward across the Pacific by tectonic plate movements, are increasingly important in explaining the region's geology. Some terranes may have transported organisms, such as the tailed frogs I described in "Running Water," whose closest relatives live in New Zealand. Originating on a South Pacific landmass in the Jurassic Period, the frogs may have spent the next one hundred million years on a biogeographical ocean liner before docking in the Northwest during the past sixty-five million.

Scientific changes have affected most of the ideas I wrote about twenty years ago. New evidence has challenged the theory I mentioned in "Running Water," that the first flowering plants evolved as shrubs along rivers. The earliest known flowering plant fossils are of small herbs. New evidence has challenged the theory I mentioned in "High Meadows" that the first bipedal hominids evolved in savanna. Some of the earliest known hominid fossils are associated with remains of forest plants, implying that hominids moved into grassland after they stood upright. New evidence has challenged the theory I mentioned in "The Evergreen Forest" that snakes originated when lizard ancestors became burrowers and lost their limbs. A 95-million-year-old Israeli fossil, of a serpentine marine creature with small hind legs, implies that snakes originated when lizard ancestors became swimmers.

Scientific changes are often inconclusive, however. Many herpetologists doubt that the Israeli fossil was a snake, since snakelike reptiles have evolved repeatedly. And even if it was, one fossil doesn't prove snakes evolved in water, any more than a few hominid bones with forest plant fossils prove hominids evolved in forest. New fossils will turn up, and theories will shift again in their kaleidoscopic way.

So many theories exist that one reader faulted me for not describing more in the book, such as punctuated equilibria, much-discussed in the 1980s, which says that evolutionary change occurs in bursts during times of environmental stress instead of continuously. It is hard to write about things you don't perceive, however, and I didn't see many bursts of evolution on my backpacking trips. I didn't see much continuous evolution either, actually. That doesn't mean evolution wasn't occurring, any more than the earth's apparent flatness means it is not a sphere. Many scientific facts are counterintuitive to human perception. The multiplicity of theories about evolution shouldn't be confused with its factuality.

A fourth kind of inaccuracy has arisen as the region itself has changed since 1983. Because of activists' work, places like the Siskiyous, Russian Peak, Red Buttes, and Rogue River Gorge, unprotected when I wrote the book, are now national forest wilderness areas, and Congress has enlarged the Kalmiopsis, Marble Mountain, Trinity Alps, and Yolla Bolly wildernesses. On the other hand, after heavy logging in the 1980s and less heavy logging in the 1990s, the book describes some vanished old growth, and such inaccuracies grow as exploiters chip away at unprotected roadless areas. Logging, mining, dams, and water diversions have pushed fish populations to new lows, and the government has listed several

Klamath Mountain taxa under the Endangered Species Act since 1983. Air pollution and other atmospheric changes are at work, although effects are less evident than in the Sierra or Cascades because the Klamaths are not downwind of big cities and agribusinesses. More visibly, increasing traffic spreads exotic organisms like a disease which kills Port Orford cedar and Pacific yew. Some watersheds that were the lacy green of cedar foliage in 1983 now have gray ghost forests.

Another exotic organism added a fatal inaccuracy to *The Klamath Knot*'s most quoted passage, a description in "Primal Ooze" of floating on a log in the Kalmiopsis's Babyfoot Lake, watching rough-skinned newts in water so clear I could see them ten feet down. The lake seemed primeval in 1979. When I went in 1998, there was a new parking lot at the trailhead, and I was puzzled to meet people with spinning rods on the trail. I learned the reason at the lake — illegally introduced smallmouth bass lurked there. I didn't find any newts, although I circled the lake, and looked again two years later. The bass wouldn't have eaten adult newts, which are highly toxic, but newt eggs and larvae evidently aren't. There may still be newts in Babyfoot Lake, but fewer than there were. Anyone wanting to explore newt consciousness must go elsewhere, at least while the bass are there.

A lot of places remain, fortunately. At one little lake above Elk Valley in the Siskiyous, newts even have turned the tables on human encroachment. I happened on the lake on a hot summer day, so I decided to jump in. Many other hikers had done so — the accessible part of the shore was worn bare. I paddled around awhile, then sat in the shallows with my lower half submerged, enjoying the amphibious ambience. Several newts sprawled nearby, and one crept to my foot and nosed against my big toe, apparently lacking the brains to go around. But it had more brains than I'd thought. I felt a tickling sensation, and looked closer. The newt was nibbling at my toe, evidently eating bits of abraded skin.

Soon others converged on me. My original attacker snapped at them, but knots of nibbling newts engulfed both feet within minutes. Still more approached, and a few swam to the surface and eyed me, as though surveying the extent of this new food source. It was unusual to see an evolutionary adaptation in action, with backpackers' toes punctuating the ancient equilibrium of newt nonanthropophagy. But I ended my observations hastily when newts headed for my crotch.

A fifth kind of inaccuracy is harder to correct than the others, because it concerns thing I'm unsure about. Some readers questioned my "evolutionary mythology," particularly my use of old-fashioned mythic elements like giants and dragons in a nature essay. One newspaper reviewer said I had "chosen to play fast and loose" with the concepts of science and mythology "in a grab for cosmic significance." Other critics liked the idea of regarding evolution as a mythology as well as a science. If scientists can theorize about the evolutionary facts, they seemed to ask, why shouldn't writers mythologize about them?

Theory and myth seem strikingly complementary at times. I wondered in "Tracks in the Wilderness" whether science ever would give people the mythic satisfaction of thinking that some landmark was where an evolutionary figure like *Brontosaurus* met its fate. Within a decade, theory produced the Yucatan meteor crater that may have ended, or helped to end, the career of sauropod dinosaurs like *Brontosaurus* (or *Apatosaurus,* its more correct, if less mythic, name). Whether or not it alone killed the dinosaurs, the Everest-sized bolide that hit the earth at the end of the Cretaceous Period would be a fitting nemesis for dragons as well dinosaurs, although it didn't kill *Apatosaurus,* which died in the Jurassic Period, tens of millions of years earlier. But perhaps theory will posit a mythically satisfying fate for *Apatosaurus* someday.

Even some friendly critics feared my giants and dragons might abet pseudosciences like creationism and paranormalism, however. Given the Gallup Poll's demonstrations that most Americans don't understand evolution's factuality, I sympathize with such fears. I never intended to oppose evolutionary science, which I find friendly and inspiring. And I admit, on re-reading the book, that the Klamaths' local giant, Bigfoot, sometimes seems to play the 400-pound gorilla in what might have been more a straightforward evolutionary tale. Still, it's one thing to read a book, and another to write it. I wrote as a backpacker sleeping in the woods, and my perceptions were not always straightforward. I didn't see Bigfoot or evidence of his existence, and I did see reasons why a nocturnal, boreal wild hominid would be an evolutionary anomaly. But I also had experiences which made me wonder about consciousness, a subject which remains mysterious, and which includes phenomena like Bigfoot sightings.

The main such experience was my sudden illness on Clear Creek in the Siskiyous. Exhaustion or a backcountry microbe may have caused it, but the mental effects were more striking than any other illness I've had. They included not only the terror and historical visions I described in the book but something I didn't. Lying in the dark, I couldn't close my eyes because intensely vivid faces would appear, mouthing incomprehensible words. The faces seemed so real that I had trouble reassuring myself that they came from my mind, and I afterward saw them at other camps, as though I'd been sensitized to something. The rational explanation was that I was sensitized to my experience in the forest, but I couldn't dismiss the possibility that I was sensitized to something *in* the forest. It made me wonder where the mind ends and the forest begins.

Unlike natural selection's co-discoverer, Alfred Russel Wallace, I don't think a supernatural power miraculously imbued "*Homo presapiens*" with a mind. I think the human mind evolved from the minds of other species. But mental evolution remains less clear than physical. Do dreams and visions evolve? We know that other mammals dream, and also that they can envision, or at least hallucinate, things. My otherwise sane cat is convinced that a feline enemy lurks permanently on the garage, and he peers upward for hours, fur bristling, at what to me is an empty roof. Maybe such things are epiphenomenal and pathological to brain function, but, in an evolutionary universe, they must have an evolutionary role, as do other epiphenomena and pathologies.

An early paleontologist named Edward Drinker Cope described an illness oddly like mine while fossil hunting in the Wyoming mountains in 1872. "I had terrible visions and dreams," he wrote, "and saw multitudes of persons, all speaking ill of me, and frustrating my attempts to sleep." Cope was a genius who discovered a surprising number of the dinosaurs and other creatures which inhabit our evolutionary stories, and whom Stephen Jay Gould, co-author of punctuated equilibria, called "America's first great evolutionary theorist." His theories seem crude now — he thought that animals could transmit acquired traits to their offspring, for example. Still, as that idea suggests, he was one of the first scientists to try to explain the role of consciousness in evolution, and his readiness to address the subject seems to have given him a grudging appreciation for old-fashioned mythologies.

Like many nineteenth century evolutionists, Cope was a racist, and when he wrote his wife from the Dakotas in 1892 about an Indian story that dinosaur bones were the remains of dragons, he was scornful. "The Sioux knew of it long ago, but they believed the bones belonged to evil monsters which were slain by lightning by the Great Spirit. They would not touch the bones for fear a like fate would befall them. So they were fortunately preserved for the more intelligent white man who is not troubled by such superstitions." Two summers of Dakota thunderstorms and eerie fossil discoveries evidently changed his attitude, however. When he repeated the story a year later, it sounded less a childish superstition and more an apt metaphor of life on the Plains — or in his own near-bankrupt laboratory in Philadelphia. "The Indians believe that the fossil bones are those of huge serpents that burrow in the earth and that lightning is always trying to find and kill them, and that those bones we see have been so killed. The bad lands have been made by the efforts of lightning to find them in the earth." It sounded, indeed, a bit like the comet theory of dinosaur extinction.

Cope's story brings up another inaccuracy I'm unsure about. In the Klamath Mountains, as in the rest of the Pacific Northwest, Bigfoot is only recently an evolutionary story. Before whites arrived and adapted him to their own uses, he was known to the people who have been there for many thousands of years. I noted this in the book, and mentioned other aspects of the original stories, but I said little more about the original people. Some readers have noted the omission, and it certainly was an inaccuracy, but not one I felt I could correct.

Native American cultures are very diverse and complex, and the Klamath region's are particularly so, as anyone knows who has read the great anthropologist Alfred Kroeber's attempts to explain them. I was aware enough of this when writing the book to feel that the subject was beyond my grasp. And there was a deeper reason why I shied away, although I was less aware of it. In using my evolutionary story to describe the Klamath Mountains, I was behaving rather as Edward Cope had in digging up Dakota dinosaurs. Indeed, although I don't regard Indians as inferior, I may have been more of a mythological encroacher than he was. The Lakotas in the 1890s probably thought Cope's ideas merely odd. By the 1980s, many tradition-minded Indians regarded evolution as a threat, and not without reason. They saw the idea that cultures evolve progressively, from hunting-and-gathering to farming-and-industry, as a denial of their own traditions' value, and they saw the idea that their ancestors evolved in the Old World before migrating here as a threat to their land rights. So they understandably might have disliked my tying them into *The Klamath Knot*. The only traditionalist who commented on the book to me was appreciative, or at least tactful, but that doesn't mean it isn't part of a five-hundred-year-old "inaccuracy."

I don't think evolution is a large part of the "inaccuracy." It came late, and has been more a rationalization for encroachment than a motive. Like racism, cultural "progressism" now is discredited as an evolutionary concept. And whether or not native cultures have occupied the Klamaths since the Earth's beginning, the fact they've done so for thousands of years, maintaining forests and fisheries in a better state than today's, is a sound claim by any moral standard.

Neither evolution nor Native American culture is going away, so adaptations will occur. A story I heard recently interests me. The Karuks, who live along the Klamath upriver from Bluff Creek, say Bigfoot originated from human misbehavior. Because Karuk villages were isolated, they had a rule that people should marry outside. A couple in one village started living together, however, and when they refused to separate, the elders expelled them. The couple moved downriver asking to live in other villages, but were refused out of respect for the elders. When they arrived at Bluff Creek, they realized that they were outcasts, so they went up the creek to live alone. Eventually, they fled into the high country, where people don't live, and evolved (the teller used the word) into large, shaggy beings who could survive the snow forest winters.

I can read several evolutionary ideas into the story. The rule against intra-village marriage is eugenic, recognizing the advantages of out-crossing. The evolution of the Karuk couple's descendants into Bigfoot is an example of allopatric speciation, the process whereby an isolated population diverges from its parent species. It is also a good demonstration of Darwinism's basic principle that natural selection is directed toward adaptation to environment. The couple's descendants get big and hairy to survive in the snow forest, not to become "more highly evolved." It is a better demonstration than the story of Bigfoot as an ancestral hominid, because that story can imply an inherent tendency for humans to get "more highly evolved," whereas Darwinism doesn't. As I wrote in the last chapter, Darwinism doesn't say humanity is *destined* for anything.

The Karuk story implies an alternate evolutionary explanation for Bigfoot sightings, although probably not one my critics will like. Human culture — language and technology — evolved during the past two million years because it was adaptive to a rich, diverse natural environment. If we continue to impoverish and simplify the planet, culture will be less of a selective advantage. If the biosphere finally is so impaired that technology and language become inadaptive, the brain will lose them, and humans, should they survive, will become big, hairy beasts like the heedless Karuk couple's descendants, because that is adaptive. Perhaps Bigfoot sighters are getting a preview of this in the Klamath night. Physics increasingly is exploring the fourth dimension, and if time travel is scientifically conceiveable, then "time vision" might be too. One exasperated reviewer said I was trying to turn the Klamaths into a "Bermuda Triangle of the mind" when I wrote of foreseeing rattlesnake encounters in "High Meadows," but, coincidentally or not, I foresaw rattlesnake encounters.

I acknowledge a problem with this explanation, however. "*Homo postsapiens*" probably wouldn't resemble the ape-like hominid of Bigfoot sightings. Evolution doesn't regress any more than it progresses, so the new species would have more traits in common with *H. sapiens* than with *H. presapiens*. It might be like a creature I dreamt about in 1984. His cranium was small, but he had an aquiline nose and a pointed chin instead of an ape's flat nose, receding chin, and brow ridges. His tan hair, including a dark beard and a mane, seemed more equine than simian, and long, thin legs gave him a spidery gait as he climbed a serpentine ridge. I was in the Yolla Bollys when I dreamt him, but nobody actually seems to have sighted such an animal.

Whatever the value of "evolutionary mythology," *The Klamath Knot*'s portrayal of Bigfoot as a big part of human perceptions of the region remains accurate. Sightings continue, despite shrinkage

of wildlands, and Al Hodgson, an unassuming, level-headed individual who made plaster casts of huge tracks at Bluff Creek in the
1960s, has started a Bigfoot museum and research institute
in Willow Creek. (A statue outside the museum looks like a hairy
version of the *Homo postsapiens* I dreamt about, but is considered
less authentic than an older, ape-like one.) Bigfoot's significance in
traditional Native American culture may be growing. At the 2000
California Wilderness Conference in Sacramento, an Indian man
stood before the assembled conservationists and remarked that he
hadn't heard any mention of Bigfoot there. He didn't elaborate, so
I don't know exactly what he meant, but I gather he thought Bigfoot should have been mentioned. If so, I agree with him. Bigfoot is
a manifestation of the unknown and unexpected, and I think we
need that if we're to avoid the exhausted world of *Homo postsapiens*.

That brings up a sixth and last kind of inaccuracy, the kind that
will occur in the future as change continues. I would be glad if the
book's references to sparse salmon runs became inaccurate because abundant salmon runs return. But references to abundant
roadbuilding, logging, and mining would have to become inaccurate for that to happen.

It won't happen unless the Klamaths get more attention. A region with the most wild rivers on the west coast and the richest
temperate coniferous forest anywhere certainly seems to deserve
it. The International Union for the Conservation of Nature classes
the Klamaths among the world's 200 most botanically significant
places. Yet they figured relatively little in mainstream conservation
agendas during the past two decades. Of the big private organiza-
tions, only the World Wildlife Fund and, to a lesser extent, the
Wilderness Society and National Wildlife Federation, showed much
interest in them on a national level. And while the Clinton administration's roadless area protection initiatives slowed the hemorrhage of North American biodiversity, a piecemeal, band-aid
quality to its efforts was particularly evident in the Klamaths. The
administration left them off its original initiative to curtail road
building, and left some important roadless areas out of its national
forest conservation plan.

Exploiters want to tear off the band-aids, and will get many
chances if history is a guide. In the 1930s, when Aldo Leopold
and Bob Marshall started administrative protection of wilderness

areas, most trailheads into Klamath Mountain backcountry were on national forest boundaries or major roads. In 1964, when Howard Zahniser finally pushed the Wilderness Act's legislative protections through Congress, Klamath trailheads were at the ends of hundreds of miles of logging roads that the Forest Service had built through former administrative wilderness. I know of only one wilderness area trailhead on a major road in the Klamaths today, at Fort Goff on State 96 west of Yreka. I entered the Red Buttes from it in 1979, and it was still there in 1999. But miles of old growth separate it from the wilderness boundary, so who knows how much longer it will be there?

The only permanent new protection in recent years came at the region's eastern edge, when President Clinton proclaimed a 52,000 acre Cascade-Siskiyou National Monument, and it wasn't Washington that envisioned that area as an ecological corridor or "loading dock" between the Klamaths and Cascades. It was local activists like Dave Willis, who cautioned that it's not enough to protect a loading dock if you neglect the biodiversity "warehouse." Clinton failed to proclaim a much larger Siskiyou Wild Rivers National Monument that local activists proposed for the Kalmiopsis area, although his Interior Secretary, Bruce Babbitt, called it "probably the most important, the most biologically significant, unprotected landscape in the American West." And it wasn't just Clinton who overlooked the proposal. Most news media ignored it. My attempts to interest environmental magazines in it failed.

Siskiyou Wild Rivers National Monument's proponents did everything they could to make it possible. They didn't propose that the government buy land, just protect what it owns in Siskiyou National Forest. They didn't even propose transferring the land from the Forest Service to the National Park Service, which traditionally manages national monuments. In the context of the past two decades, they did the sensible thing. But I think that context has lost touch with something.

Conservation increasingly seems an art of the known and expected. We know that giant sequoias and redwoods can be saved, for example, because of Sequoia and Redwood national parks. So the Clinton administration saved more of them in Sequoia National Monument and Headwaters Reserve. But national parks and wilderness areas are part of an "American idea" that was unknown and unexpected during civilization's first 5,000 years of tyranny and land rape. "If the stars did not dance in the sky when our Constitutional Convention met," wrote a conservation-minded historian, Bernard de Voto, "I don't know what romance is. Ours is a story mad with the impossible." Ferdinand Hayden, Teddy Roosevelt, and John Muir weren't practicing "the art of the possible" when they established millions of acres of national parks, monuments, and wildlife refuges. Maybe it's time to start pursuing the unexpected again.

The Klamaths would be a good place to begin. We *don't* know if Port Orford cedars can be saved. The region should be a biosphere reserve, like its deciduous forest counterpart, the Southern Appalachians. It amply deserves that status for protection and restoration of its forest, fisheries, and indigenous cultures. Local activists have supported the idea for decades. But biosphere reserves are supposed to center on a national park or other "core reserve" like Great Smokies in the Appalachians, and a new national park in the Klamaths would be a surprise. As of 2002, the only national park system unit in the New Hampshire-sized region is 480-acre Oregon Caves National Monument, proclaimed in 1909. It is one of the nation's smallest units, and politics stymied a 1998 Park Service proposal to enlarge it by 3,410 acres.

A national park wouldn't protect the whole region, of course, but it would help. The question is, where would it be? The Klamaths' big wilderness areas have great park potential, but I think making one of them a park would be more of the known and expected. They are the grandest areas, and national parks in the West need to go beyond grandeur. Also, each area has its own particular character that is not quite representative of the whole region. So I would leave them to Bigfoot, although old growth and riparian corridors for wildlife migration should link them much more than at present. California and Oregon wilderness bills introduced in Congress in 2002 would help accomplish that.

The far West is getting as trampled as the East, so a new national park in the Klamaths might take inspiration from Mammoth Cave National Park in Kentucky. It also contains a global superlative — the world's largest cave system. When the park began in the 1930s, the caves had been mines and tourist traps for over a century, while the hills and streams were deforested and polluted.

Today it has one of the largest areas of lowland deciduous forest in the southeast. The ecosystem has not recovered completely — a bat population originally numbering millions remains in thousands. But the park is surprisingly alive. One cave, notorious when a trapped commercial caver named Floyd Collins died there, is an example. A 1925 photo made during the rescue attempt shows bare dirt with concession stands and crowds. Today, big beeches, maples, and tulip trees surround the entrance, and no hint of the carnival past remains. It's one of the quieter places I've experienced. Yet this exists in a park where hundreds tour the caves every day.

I hadn't known what surprising habitats big caves are until I went on a Mammoth tour. There's nothing like walking underground through miles and miles and *miles* of Paleozoic rocks to get a sense of geological deep time. A Klamath Mountain park could provide hundreds of people daily with a corresponding sense of biological deep time. It could become the global flagship of temperate coniferous forest, as Mammoth Cave is for limestone caverns. To do so, however, it would need what Mammoth Cave has — a good visitor infrastructure — and what Mammoth hasn't — an arboretum-museum where the general public could see how the forest evolved and how it works. It would need "genes" from the Smithsonian and the Arizona-Sonora Desert Museum as well as Mammoth Cave.

Such a park would be hard to grow, but seeds exist near the region's center, in the area encompassing Oregon Caves National Monument and Red Buttes Wilderness. This is a kind of hologram of the Klamath Mountains, with glaciated metamorphic peaks, tarns and meadows, serpentine ridges, marble caverns, and deep gorges of ancient forest. Some of the region's best tree specimens grow there. It also borders the Klamath River, which remains the region's definitive one although upstream dams and diversions have diminished it.

The area typifies neglected and failed aspects of national conservation in the West. Oregon Caves was managed more as a tourist attraction than a natural area for most of its existence, and is much, much too small. Red Buttes is classic "wilderness on the rocks," a clearcut-bound spine of peaks that falls far short of Aldo Leopold's two-week pack trip criterion for wilderness size. Both units should be enlarged, and linking them in a park would do that.

They also could be linked to Cascade-Siskiyou National Monument, providing something for the "loading dock" to load.

Such links might help to correct another failure of western conservation. Crater Lake National Park in the Cascades northeast of the Klamaths is a textbook example. It is a grand place, one of the continent's deepest, clearest lakes, created by one of the greatest volcanic explosions. But its history demonstrates the pitfalls of protecting land just for its grandeur. When the park was established in 1902, its fauna included a number of native mammal species, including otters, ermine, mink, and spotted skunks as well as wolves and grizzlies. None of these were known to occur there when a researcher visited it in the 1980s. They had vanished because the park lacked enough habitat, and logging and other impacts had destroyed it in surrounding areas. A Klamath Mountain park linked with wilderness areas and riparian corridors might gain species instead of losing them.

The park would protect wildlands, but it would also show the value of temperate coniferous forest biodiversity to as many people as possible. It would be accessible, since it is close to the region's three main transportation corridors. It would be easy (or at least inexpensive) to establish, since the southern unit of Rogue River National Forest would provide most of the land, with additions from Siskiyou and Klamath national forests. Much of the land around Oregon Caves and Red Buttes, among the most heavily logged and roaded in the region, would have to be restored, but that is a popular process that makes jobs and demonstrates biodiversity management. The park would restore fisheries as well as forest. There should be a wild place outside Alaska where a lot of people can see a lot of anadromous fish, maybe lampreys and sturgeon as well as salmon and steelheads.

A new park including Oregon Caves and Red Buttes could mitigate one of the Klamath Mountains' conservation liabilities. Two-thirds of the region is in California, but most park proposals have been in Oregon. A proposal in both states might magnify support. On the other hand, California already has six national parks to Oregon's one, and establishing interstate conservation units has always been hard, not least if the states are Oregon and California. The Cascade-Siskiyou National Monument proposal included land in both. The existing monument is in Oregon.

The state boundary is just one problem, of course. I know a new national park in the Klamath Mountains is a daydream in the present political landscape. An unexpected and unknown continent has shrunk to a five-hour jet ride, and deadlocked factional agendas smother what Wallace Stegner called "the geography of hope." But I don't mind daydreaming about an unexpected and unknown Klamath Mountains, because civilization's fantasy of a known and expected world is idler. Who could have expected that life is four billion years old, and who knows what it will be in another million? Maps still have dragons on their edges, although we call them quarks and strings. I don't know if a cosmic string is less metaphorical than a dragon. I do know that an evolving universe recedes from knowledge and expectation at the speed of light, and will continue no matter how fast and far pursued. As a scientist nine centuries ago wrote:

Up from earth's center, through the seventh gate
I rose, and on the throne of Saturn sate,
And many knots unraveled by the road;
But not the knot of human death and fate.

Rubaiyat

Azoic era Eras are the largest segments of geological time. The Azoic, meaning "without life," is the earliest, extending from the earth's beginnings to the first evidence of life. 25, 32

Cretaceous period Eras are divided into periods. The Cretaceous, which extended from about 136 million years ago to about 65 million years ago, was the last period of the Mesozoic era. 24, 57, 128

Devonian period The Devonian lasted from about 395 million years ago to about 345 million years ago, during the Paleozoic era. 41, 45

Eocene epoch Periods are divided into epochs. The Eocene was the second epoch of the first period of the most recent era (the Cenozoic era). It lasted from about 54 million years ago to about 38 million years ago. 6

Jurassic period The Jurassic preceded the Cretaceous during the Mesozoic era, extending from about 190 million years ago to 136 million years ago. 24, 46

Mesozoic era The Mesozoic, meaning "middle life," came between the Paleozoic and Cenozoic eras and lasted from about 225 million years ago to 65 million years ago. 23

Miocene epoch The Miocene lasted from about 26 million years ago to 6.3 million years ago. 117

Oligocene epoch The Oligocene preceded the Miocene, lasting from about 38 million years ago to 26 million years ago. 25, 112, 117

Paleogene period The Paleogene was the first period of the Cenozoic era. Also called the Tertiary, this period lasted from the end of the Cretaceous period to the end of the Miocene epoch. 24

Paleozoic era The Paleozoic, meaning "ancient life," lasted from about 570 million years ago to about 225 million years ago. 13, 23, 27, 39, 48, 51, 128

Pleistocene epoch The Pleistocene lasted from about 2 million years ago until the present epoch, the Holocene. 25, 117, 119, 123

Pliocene epoch The Pliocene preceded the Pleistocene, beginning about 6.3 million years ago. 5, 117, 127

Actinomycete, order Actinomycetales. 35

Alga, blue green, phylum Cyanophyta. 33, 34, 35

Alga, brown, division Phaeophyta. 58

Alga, diatom, division Chrysophyta. 37

Alga, green, division Chlorophyta. 37, 58

Amphipod, order Amphipoda. Scud. 40

Aplodontia, *Aplodontia rufa*. 93

Aspen, *Populus tremuloides*. Quaking aspen. 97, 115

Aster, genus *Aster*. 117

Azalea, *Rhododendron occidentale*. Western azalea. 4, 29, 60, 78

Bacterium, phylum Schizomycophyta. 33, 34, 35

Badger, *Taxidea taxus*. American badger. 122

Bald cypress, genus *Taxodium*. 25, 77

Baneberry, *Actaea rubra*. Western baneberry. 93

Bay laurel, *Umbellularia californica*. California bay, Oregon myrtle, pepperwood. 5, 78

Bear, black, *Ursus americanus*. 3, 4, 93, 120

Bear, grizzly, *Ursus arctos*. 4, 119

Bear grass, *Xerophyllum tenax*. 97

Beaver, *Castor canadensis*. 14

Beech, genus *Fagus*. 6

Bitter cherry, *Prunus emarginata*. 97

Blackberry, genus *Rubus*. 78

Bladderwort, genus *Utricularia*. 48

Blueberry, genus *Vaccinium*. 78

Blue-eyed Mary, genus *Collinsia*. 88

Bobcat, *Lynx rufus*. 103

Bog laurel, *Kalmia polifolia*. Alpine laurel, American laurel. 29

Buckeye, *Aesculus californica*, California buckeye. 78

Butterfly, California tortoiseshell, *Nymphalis californica*. 104

Cascara, *Rhamnus purshiana*. Cascara sagrada, cascara buckthorn, 78

Ceanothus, genus *Ceanothus*. 78, 104

Cedar, incense, *Libocedrus decurrens*. 4, 78, 96, 108

Cedar, Port Orford, *Chamaecyparis lawsoniana*. 4, 78, 107

Centipede, class Chilopoda. 85

Chestnut, genus *Castanea*. 77

Frog, tailed, *Ascaphus truei*. 63-65, 87
Frog, yellow-legged, *Rana boylei*. 63
Garter snake, genus *Thamnophis*. 91
Gentian, genus *Gentiana*. 117
Ginkgo, genus *Ginkgo*. 6
Golden chinquapin, *Castanopsis chrysophylla*. 4, 5, 76, 96
Gooseberry, genus *Ribes*. 78
Grass, big bluestem, *Andropogon gerardii*. 119
Grass, bunch, genus *Stipa, Danthonia, Festuca, Poa,* etc. 88, 110
Grass, cheat, *Bromus secalinus*. 110
Grosbeak, black-headed, *Pheucticus melanocephalus*. 87, 88, 91, 94
Ground cone, *Boschniaka strobilacea*. 82-83, 102
Ground squirrel, California, *Citellus beecheyi*. 120-121
Ground squirrel, golden-mantled, *Citellus lateralis*. 101
Grouse, blue, *Dendragapus obscurus*. 88, 91-92
Grouse, ruffed, *Bonasa umbellus*. 88
Hawk, red-tailed, *Buteo jamaicensis*. 121
Hawthorn, genus *Crataegus*. 78
Hazel, *Corylus cornuta*. 78
Heather, *Phyllodoce empetriformis*. Red mountain heather (or *Cassiope mertensia,* white heather). 4
Hemlock, mountain, *Tsuga mertensia*. 78, 97
Hemlock, western, *Tsuga heterophylla*. 78
Hermit thrush, *Hylocichla guttata*. 90, 94, 108
Hickory, genus *Carya*. 6
Holly, genus *Ilex*. 77
Horsetail, genus *Equisetum*. 43, 84, 113
Huckleberry, genus *Vaccinium*. 78
Hummingbird, calliope, *Stellula calliope*. 123
Hydra, genus *Hydra*. 38
Inside-out flower, *Vancouveria hexandra*. 93
Jackrabbit, black-tailed, *Lepus californicus*. 120
King snake, California mountain, *Lampropeltis zonata*. 91
Klamath plum, *Prunus subcordata*. 97
Lamprey, Pacific, *Entosphenus tridentatus*. 56-57
Larkspur, genus *Delphineum*. 117
Lily, leopard, *Lilium pardalinum*. 29, 118
Lizard, alligator, genus *Gerrhonotus*. 88
Lizard, fence, *Sceloporus occidentalis*. Western fence lizard. 88, 112

BIBLIOGRAPHY

Axelrod, Daniel I. *History of the Coniferous Forests, California and Nevada.* University of California Publications in Botany. Vol. 70. Berkeley and Los Angeles: University of California Press, 1976.

Banks, Harlan P. *Evolution and Plants of the Past.* Belmont, California: Wadsworth, Inc., 1970.

Barbour, Michael G., and Major, Jack, eds. *Terrestrial Vegetation of California.* New York: John Wiley & Sons, 1977.

Beck, Charles B., ed. *Origin and Early Evolution of Angiosperms.* New York and London: Columbia University Press, 1976.

Burton, Maurice. *Living Fossils.* London: Thames and Hudson, 1954.

California Department of Fish and Game. *Trout of California.* Sacramento, 1969.

Campbell, Joseph. *The Masks of God: Creative Mythology.* London: Secker and Warburg, 1968.

_____. *The Masks of God: Occidental Mythology.* London: Secker and Warburg, 1965.

_____. *The Masks of God: Oriental Mythology.* London: Secker and Warburg, 1962.

_____. *The Masks of God: Primitive Mythology.* New York: The Viking Press, 1959.

Colbert, Edwin H. *Evolution of the Vertebrates: A History of the Backboned Animals through Time.* New York: John Wiley & Sons, 1955-1969.

Cowen, Richard. *History of Life.* New York: McGraw-Hill Book Co., 1976.

Darrah, William C. *Textbook of Paleobotany.* New York: D. Appleton-Century Co., 1939.

Dillon, Richard. *Siskiyou Trail: The Hudson's Bay Fur Company Route to California.* New York: McGraw-Hill Book Co., 1975.

Downs, Theodore. *Fossil Vertebrates of Southern California.* Berkeley and Los Angeles: University of California Press, 1968.

Fraser, Sir James. *The New Golden Bough.* Revised and edited by Theodor H. Gaster. New York: Criterion Books, 1959.

Gardner, John. *Grendel.* New York: Alfred A. Knopf, 1971.

Gayley, Charles Mills. *The Classic Myths in English Literature and Art.* Lexington, Massachusetts: Xerox College Publishing, 1939.

Gould, Stephen Jay. *Ever Since Darwin: Reflections in Natural History*. New York: W.W. Norton, 1977.

_____. *The Panda's Thumb: More Reflections in Natural History*. New York: W.W. Norton, 1980.

Green, John. *Bigfoot: On the Track of the Sasquatch*. New York: Ballantine Books, 1973.

_____. *Sasquatch: The Apes Among Us*. Seattle: Hancock House Publishers, Inc., 1978.

Haig-Brown, Roderick L. *Return to the River: A Story of the Chinook Run*. New York: William Morrow & Co., 1941.

Hart, John. *Hiking the Bigfoot Country: Exploring the Wildlands of Northern California and Southern Oregon*. San Francisco: Sierra Club Books, 1975.

Howard, Hildegarde. *Fossil Birds*. Los Angeles County Museum Science Series no. 17, Paleontology no. 10, February 1962.

Irwin, William P. *Geology of the Klamath Mountains Province*. Bulletin 190, California Division of Mines and Geology, Ferry Building, San Francisco, 1966.

Kozloff, Eugene. *Plants and Animals of the Pacific Northwest*. Seattle: University of Washington Press, 1976.

Lanham, Url. *The Fishes*. New York: Columbia University Press, 1962.

Ley, Willy. *Dragons in Amber*. New York: The Viking Press, 1951.

_____. *Salamanders and Other Wonders*. New York: The Viking Press, 1955.

Lloyd, Francis Ernest. *The Carnivorous Plants*. New York: Dover Publications, 1976.

McKee, Bates. *Cascadia: The Geological Evolution of the Pacific Northwest*. New York: McGraw-Hill Book Co., 1972.

Moody, Richard. *The Fossil World*. Secaucus, New Jersey: Chartwell Books, 1977.

Moyle, Peter B. *Inland Fishes of California*. Berkeley and Los Angeles: University of California Press, 1976.

Netboy, Anthony. *The Salmon: Their Fight for Survival*. Boston: Houghton Mifflin, 1974.

Niehaus, Theodore F. and Ripper, Charles L. *A Field Guide to Pacific States Wildflowers*. Boston: Houghton Mifflin, 1976.

Norris, Robert M., and Webb, Robert W. *Geology of California*. New York: John Wiley & Sons, 1976.

Oakeshott, Gordon B. *California's Changing Landscape: A Guide to the Geology of the State.* New York: McGraw-Hill Book Co., 1971.

Pennak, Robert W. *Freshwater Invertebrates of the United States.* New York: The Ronald Press Company, 1953.

Raven, Peter H., and Axelrod, Daniel I. *Origin and Relationships of the California Flora.* University of California Publications in Botany, vol. 72. Berkeley and Los Angeles: University of California Press, 1978.

Rhodes, Frank H.T.; Zim, Herbert S.; and Schaffer, Paul R. *Fossils: A Guide to Prehistoric Life.* New York: Golden Press, 1962.

Sagan, Carl. *The Dragons of Eden.* New York: Random House, 1977.

Sanderson, Ivan T. *Abominable Snowmen: Legend Come to Life.* Philadelphia and New York: Chilton Book Company, 1961.

Scientific American. *Continents Adrift and Continents Aground: Readings from Scientific American.* San Francisco: W.H. Freeman & Co., 1976.

——————. *Continents Adrift: Readings from Scientific American.* San Francisco: W.H. Freeman & Co., 1970.

Seward, A.C. *Plant Life Through the Ages.* London: Cambridge University Press, 1941.

Simpson, George Gaylord. *The Major Features of Evolution.* New York: Simon and Schuster, n.d.

Smith, Gilbert M. *Freshwater Algae of the United States.* New York: McGraw-Hill Book Co., 1950.

Stebbins, Robert C. *Amphibians and Reptiles of Western North America.* New York: McGraw-Hill Book Co., 1954.

Tidwell, William G. *Common Fossil Plants of Western North America.* Provo, Utah: Brigham Young University, 1975.

Usinger, Robert L., ed. *Aquatic Insects of California.* Berkeley and Los Angeles: University of California Press, 1963.

Warburton, Austen D., and Endert, Joseph F. *Indian Lore of the North California Coast.* Santa Clara: Pacific Pueblo Press, 1966.

Whittaker, R.H. *The Ecology of Serpentine Soils.* Ecology, 35: 275-288, 1954.

——————. *Vegetation of the Siskiyou Mountains.* Ecological Monographs 30:279-338.